Social Responsibility in Science, Technology, and Medicine

Social Responsibility in Science, Technology, and Medicine

Paul T. Durbin

Bethlehem: Lehigh University Press
London and Toronto: Associated University Presses

Associated University Presses
440 Forsgate Drive
Cranbury, NJ 08512

Associated University Presses
25 Sicilian Avenue
London WC1A 2QH, England

Associated University Presses
P.O. Box 39, Clarkson Pstl. Stn.
Mississauga, Ontario,
L5J 3X9 Canada

The paper used in this publication meets the requirements of the American National Standard for Permanence of Paper for Printed Library Materials Z39.48-1984.

Library of Congress Cataloging-in-Publication Data

Durbin, Paul T.
Social responsibility in science, technology, and medicine / Paul T. Durbin.
p. cm.
Includes index.
ISBN 0-934223-27-0 (alk. paper)
1. Science--Social aspects. 2. Technology--Social aspects.
3. Medicine--Social aspects. I. Title.
Q175.5.D85 1992
303.48'3--dc20 92-72013
CIP

PRINTED IN THE UNITED STATES OF AMERICA

Contents

Preface

Since 1974, it seems safe to guess, well over a million people have read Robert Pirsig's *Zen and the Art of Motorcycle Maintenance*. However, it seems equally likely that only a small fraction of those readers—people like myself who reached our intellectual maturity in the heady days of the anti-Vietnam War *cum* antitechnology movement of the late 1960s and early 1970s—really appreciate the intellectual ferment out of which that remarkable book emerged.

Like others, I was impressed by the fulminations of Jacques Ellul and Herbert Marcuse against our technocratic and bureaucratic culture. Perhaps unlike others, I have taken a very long time—twenty years—to get my feelings about our technological society straightened out and to get my reactions down on paper. This book is one philosopher's personal reaction to the Great Technology Debate. I claim no great originality for the effort; it is simply my attempt to come to grips with what I see as the exaggerated claims of Ellul and Marcuse—and Martin Heidegger or literary antitechnologists such as Kurt Vonnegut, Jr. It is my attempt to get at "the truth" (my truth, with a small "t") about technology and our technological culture. I hope, however, that the book is definitive in one sense: I want it to be my last word, my last contribution to the technology debate of the 1970s. I have no hope that the book will be definitive in any other sense—though I would be naive if I did not admit that I hope others share my modest perspective.

Technology has brought many problems to our age and promises to bring still more in the future, but technology is not a force that cannot be controlled by democratic means . . . *if* enough people have the spirit, energy, and drive to mobilize against technosocial ills. My answer to the antitechnology pessimists is a modest and limited message of hope.

I have been aided in my search for this "middle path" in the technology debate by a great many friends and colleagues. To thank each of them individually would take a long, long time—though the task would be a delight rather than a chore. Instead I will thank, collectively, all the wonderful people who have worked with me on four projects: (1) the editing of the *Research in Philosophy and Technology* and *Philosophy and Technology* series for the Society for Philosophy and Technology; (2) the editing of *A Guide to the Culture of Science, Technology, and Medicine*, under the auspices of the Ethics and Values Studies Program at the National Science Foundation and the Humanities,

Science, and Technology Program at the National Endowment for the Humanities; (3) under the same auspices, a Sustained Development Award, in which NSF and NEH allowed me to do "missionary" work at my own University of Delaware, attempting to spread the ethics-and-values word to our Colleges of Engineering, Human Resources, Nursing, and Urban Affairs; it was while working under that award that I completed some of the basic research for this book; and (4) the institutionalization of the National Association for Science, Technology, and Society, on whose board I feel privileged to serve. The hundreds of people involved with those four projects will know who I mean; I hope they will accept this collective note as a heartfelt individual word of thanks to each.

I would like to dedicate the book to the hundreds—or is it thousands?—of friends and colleagues I have known over the years who have devoted so much of their energy to progressive activism, especially in terms of ameliorating problems associated with science and technology.

In this regard, I was tempted to write, "to friends in the progressive movement." But that would not capture what I have in mind, for two reasons. Talk of "the movement" might suggest people more radical than those to whom I want to dedicate the book. I have many radical friends, too, but they are not likely to resonate to my message here. Second, "the movement," even if qualified with the adjective "progressive," suggests something much more definite than is the case. What I hope, in fact, is to motivate people, currently working diligently on *separate* progressive causes, to come together to form a new progressive movement that *can* get something done about technosocial problems.

I must also thank our Philosophy Department secretaries, Mary Imperatore and Gail Ross, for their invaluable help; indeed, indispensable would be a better word, since without their help the book would never have gotten done. I also thank, for their encouragement, my wife, Lydia, my children, Mary, Maggie, Michael, and Angela; and my stepchildren, Todd and Amy Elliott.

Publication acknowledgments: Early versions of a number of chapters or parts of chapters have appeared in print before or will appear later in forthcoming proceedings volumes. I wish to acknowledge these as follows:

Chapter 1: The first part is expected to appear, under the title, "Culture and Technical Responsibility," in *New Worlds, New Technologies, New Issues*, ed. S. Cutcliffe (Bethlehem, Pa.: Lehigh University Press, 1992), 104-116. Another section (in expanded form) appears as a review of Larry Hickman's *John Dewey's Pragmatic Technology* in *Research in Philosophy and Technology*, vol. 12, ed. F. Ferré (Greenwich, Conn.: JAI Press, 1992), 347-351.

Chapter 2: An early draft appeared as "Ethics as Social Problem Solving: A Mead-Dewey Approach to Technosocial Problems," in *Evaluer la technique: Aspects éthiques de la philosophie de la technique*, ed. G. Hottois (Paris: Vrin, 1988), 161-173.

Chapter 7: This appeared, with only minor differences, as "Bioengineering, Scientific Activism, and Philosophical Bridges," in *Bridges: An Interdisciplinary Journal of Theology, Philosophy, History, and Science* 2(Spring/Summer1990): 27-41.

Chapter 10: This appears, again with only minor modifications, as "Environmental Ethics and Environmental Activism," in *Research in Philosophy and Technology*, vol. 12, ed. F. Ferré (Greenwich, Conn.: JAI Press, 1992), 107-117.

Chapter 11: An early version of most of this chapter was presented at a Society for Philosophy and Technology conference in Bordeaux, France, in 1989; it should appear, as "Technology and Democracy," in *Democracy in a Technological Society*, ed. L. Winner (Dordrecht: Kluwer, forthcoming).

Chapter 12: Early material for this chapter appeared in two book reviews: one, of Langdon Winner's *Autonomous Technology*, in *Social Indicators Research* 7 (1980):495-504; the other, of Albert Borgmann's *Technology and the Character of Contemporary Life*, in *Man and World* 2 (1988):231-239.

Chapter 13: A significant portion of this chapter appears as a review of Edmund Byrne's *Work, Inc.* in *Research in Philosophy and Technology*, vol. 12, ed. F. Ferré (Greenwich, Conn.: JAI Press, 1992), 356-360.

Chapter 16: Early material for one part of this chapter appeared as "Research and Development from the Viewpoint of Social Philosophy," in *Technological Transformation: Contextual and Conceptual Implications*, eds. E. Byrne and J. Pitt (Dordrecht: Kluwer, 1989), 33-45. Material for another part of the chapter was presented as a paper at the Frontiers in American Philosophy conference, at Texas A&M University in 1988, and should appear in the proceedings volume for that conference.

Part I

Introduction and Orientation

1

Introduction: The Crisis of Technological Culture and a Possible Solution

> *What I relate is the history of the next two centuries. I describe what is coming, what can no longer come differently: the advent of nihilism. . . . For some time now our whole European culture has been moving as toward a catastrophe, with a tortured tension that is growing from decade to decade: restlessly, violently, headlong, like a river that wants to reach the end, that no longer reflects, that is afraid to reflect.*
>
> *—Friedrich Nietzsche*

I begin with Daniel Bell quoting Friedrich Nietzsche[1]: the modern world, with technology as its driving force, has reached its culmination in cultural nihilism. I am much more optimistic than Nietzsche. I hope we can create new meaningful symbols, a new culture for our troubled technological world. I think we can do this, however, only if we struggle to do so.

I use this concern over technology and culture to launch into my main project here,[2] which is to deal with social responsibilities of scientists, engineers, medical researchers, and other technical experts. I believe that to exercise their responsibilities they must work alongside other activists trying to make ours a better world. And a better world is not just a world in which the great social problems of the day have been solved; it is one in which as many human beings as possible find a fuller meaning for their lives—find new cultural symbols that celebrate the significance of new accomplishments at the same time that they motivate the communities in which we live to continue the perpetual struggle to seek ever better worlds.

In general, I take culture to mean those parts of social behavior that have to do with what makes life worthwhile in any social group; that is, culture refers not to just any activities but to those that have an enhanced quality. More specifically, these aspects of social behavior tend to be institutionalized as the whole range of the fine arts (popular arts tend to get included only insofar as they represent something of enduring value), higher learning and its associated professions (law, for instance, tends to be included not in its mundane but in its more "philosophical" dimensions), education more broadly, and such leisure pursuits as tourism and the enjoyment of nature. In my view, furthermore, religion and religious institutions definitely fall within the scope of culture insofar as they claim to provide a spiritual meaning for life.

THE PROBLEM

Bell on the Nihilism of Modernism under Technocapitalism

I begin with Bell for three reasons. First, because his critique of technological culture is well known. Second, because he begins with Nietzsche who has become an international symbol of the nihilism of modern culture.[3] And third, because Bell has a comprehensive view of the meaning of culture. He includes under the meaning of that term not only the fine arts but also religion and anything else that gives meaning to the life of a society in a particular historical epoch.

According to Bell, Nietzsche's is but one of two nihilistic views of our technological age. The other author he refers to is Joseph Conrad, especially in Conrad's novel, *The Secret Agent*.[4] There, Bell notes, Conrad has the First Secretary of the Russian Embassy plot an attack that "must have all the shocking senselessness of gratuitous blasphemy"; it "must be against learning—science." And, Bell summarizes for us, "The act is to blow up the Greenwich Observatory, the First Meridian, the demarcation of time zones—the destruction of time and, symbolically, of history as well."[5]

It is against this backdrop that Bell sets his project in *The Cultural Contradictions of Capitalism*:

> Is this our fate—nihilism as the logic of technological rationality or nihilism as the end product of the cultural impulses to strike down all conventions? . . . I wish to reject these seductive, and simple, formulations, and propose instead a more complex and empirically testable sociological argument.[6]

Bell's argument, he says, "stands in a dialectical relation with" his earlier book, *The Coming of Post-Industrial Society*.[7] Each concentrates on the structure of postmodern society, with one emphasizing the world of work, the other, leisure activities.[8] Bell's sociological analysis opposes holistic views of historical development as well as both functionalists (he lists Emile Durkheim[9] and Talcott Parsons[10]) and Marxists. He accuses both of having "unified" systems (a common value system or a determinate culture/economics dialectic). His view distinguishes, as an analytical framework, three axes:

(1) the techno-economic order (the "axial principle" is functional rationality, with utility as the basic value; its structure is bureaucratic and hierarchical);

(2) the polity (its principle is legitimacy, and in democracies that means the consent of the governed, with equality as the basic value in all spheres; its structure is participation *via* representation); and

(3) culture (following Ernst Cassirer, Bell says there is no unambiguous principle of change and little structure today in this realm of expressive symbolism, but a perennial return to basic existential concerns of loyalty, love, tragedy, death—and the basic value is meaningful life).

As between these, there are different rhythms of change, and there is no determinate relation of one to another. (Students of Marx and Parsons will recognize this as aimed at them.) "Sociological history," in Bell's hands, traces the "disjunction between social structure and culture," with a primary focus on a "modernism" of "rampant individualism" especially in the cultural realm.[11] Indeed, the book, though it frequently replays the expert bureaucracy themes of *The Coming of Post-Industrial Society*, is in large part a documentation of the anarchy of cultural modernism in the twentieth century—and increasingly in recent postmodernist decades.

Bell's basic point may perhaps be captured in a contrast:

> In the past, human societies have been prepared for calamity by the anchorages that were rooted in experience yet provided some transtemporal conception of reality. Traditionally, this anchorage was religion. . . . Modern societies have substituted utopia for religion—utopia not as a transcendental ideal, but one to be realized through history (progress, rationality, science) with the nutrients of technology and the midwifery of revolution.
>
> The real problem of *modernity* is the problem of belief. To use an unfashionable term, it is a spiritual crisis.[12]

To his credit, Bell admits that religion—though it might restore "the continuity of generations"—"cannot be socially manufactured"; neither can a "cultural revolution be engineered."[13] Cultural values, whether religious or artistic, arise spontaneously, not on governmental or corporate demand.

This is an impressive complement to an impressive sociological analysis of contemporary expertism and technological bureaucracy that Bell had provided earlier. (The view is now commonplace and had been elaborated still earlier by prescient authors such as John Kenneth Galbraith.[14]) As I will note in a moment, I agree that there is a great deal of anarchy in the cultural realm today.

Nonetheless, I think there are problems with Bell's analysis. For one thing, though he blanches at being called a "neo-conservative," he invites such charges. He says, over and over again, that he "respects tradition" or "authoritative judgments" about the qualities of works of art.[15] I have already noted his high regard for traditional religion. What I think he means, in rejecting the neo-conservative label, is that he does not want to return to the past. He wants new values to emerge in our own time—something with which I agree. Yet he gives very little guidance on how to go about discovering such new values.

Second, Bell uses the term "liberal" in much too homogeneous a way. I claim that liberalism has a progressive wing that Bell almost never mentions—or else he disparages what he takes to be the excessive egalitarianism of its policies. Perhaps burned by the New Left activism he so strenuously opposed, Bell can recognize little value in progressive liberal activism, and he sounds extremely pessimistic about our cultural crisis today.

Marcuse's "Aesthetic Dimension"

We might wonder if someone at the opposite end of the political spectrum might not be more optimistic. But radicals of the left do not often talk about culture in Bell's broad sense, and those who do, when they discuss what passes for culture today, are as pessimistic as Bell. This is especially true of Herbert Marcuse, generally acknowledged as the philosopher of the New Left (at least in the United States).[16]

In his excellent intellectual biography of Marcuse,[17] Morton Schoolman notes how Marcuse for a short time continued, even after the decline of the New Left in the early 1970s in the United States, to argue that "liberal strategies are best suited to push toward the realization of revolutionary goals."[18] Marcuse even picked out another change agent to replace the now-quiescent students; he said the women's liberation movement is "perhaps the most important and potentially the most radical movement that we have."[19]

However, even an optimistic Marcuse—according to Schoolman —eventually became disillusioned. In his last book, *The Aesthetic Dimension* (published in 1978; Marcuse died in 1979), Marcuse, says Schoolman, "departs from the optimism of *Counterrevolution and Revolt* and returns to the theory of technological domination and one-dimensional society."[20] In Marcuse's words: "In the present, the subject to which authentic art appeals is socially anonymous." Again: "The encounter with the truth of art happens in the estranging language and images which makes perceptible, visible, and audible that which is no longer . . . perceived, said, and heard in everyday life."[21] Schoolman concludes: "Bourgeois art [especially of the nineteenth-century romantics] is the only remaining sphere of criticism in a society wherein all forms of political and cultural discourse have bowed to the domination of technical reason."[22]

On one hand, this may seem high praise for traditional culture; it is the sole and unique source of revolutionary criticism of society left in a technological world. On the other hand—most obviously and most painfully—there is no one left today who can appreciate this critical potential of the great cultural moments of our past. At least that is what Marcuse says.

My reaction to Marcuse's pessimism is complex. I agree that much that passes for art today lacks the critical potential of nineteenth-century art. Too much of it seems faddish and commercially motivated—aimed at achieving effect and success of the sort that will get it placed in contemporary art museums or in the "new works" slot on symphony concert programs, designed to "challenge" ordinary museum or concert audiences. Too little (for my taste) seems aimed at depicting grandeur and majesty or evoking profound and universal emotional responses.

At the same time, I also find myself disagreeing. It seems to me that both Bell and Marcuse are mesmerized by critics, by mainstream museum and gallery art—and corresponding faddish music and dance and "cultural" film-making. These are what make it into "art news" media of various kinds in a commercialized world that favors a new model every year. But they are not all there is to even the fine arts; much creativity in the arts never comes to the critics' attention. And besides, the fine arts are not by any means all there is to culture. Many ordinary citizens, at every level of the socioeconomic scale, find meaning in their lives and symbols to express that meaning. This may often be in terms of a popular culture looked down upon by devotees of high culture, but that does not diminish the sense of meaning that ordinary people find in these pursuits.

Note: This is not true in all cases, and there is one area of culture that does seem to be in grave danger of extinction as a direct result of technological development. I have in mind the traditional cultures of Third World countries threatened by so-called technology transfer. I am not an expert on technology transfer to Third World countries, and my knowledge of the impact of technological development on traditional cultures in those countries is even more fragmentary—based on nothing more than anecdotes related to me by students, on novels and movies, and on an occasional published account. I do not like over-romanticizing of tribal cultures, but I believe there are wonderful manifestations of the human spirit that once were to be found in tribal culture and that run the risk, as a result of technological development, of being turned into little more than curiosities in anthropological or craft museums. I will not dwell on this matter, but I will cite two brief examples here.

Cheryl Bentsen's *Maasai Days*[23] depicts in a dramatic way the impact of Western development on one of the most romanticized of tribal cultures, that of the Masai in Kenya. Bentsen is concerned especially with how young people, caught between no-longer-viable traditional ways and still-developing new social patterns, attempt to find meaning in their lives even though they are almost totally disaffected from their own people in the tribal camps.

Another example is the Bushmen in Botswana whose simple and peaceful ways are humorously contrasted with civilized life in the movie, *The Gods Must Be Crazy*. I have seen reports[24] recently which say that over the past two decades or so the government has dug wells and the Bushmen have become settlers around those wells. One of the major cultural dislocations seems to have been that tribal authority has been undermined, and disputes formerly handled at the local level are now supposed to be dealt with in modern magistrate's courts in the capital (much less successfully, it seems).

Whether the Masai or the Bushmen are better off or not—say, in terms of health or other alleged good effects of modernization—is not the issue. The point I am making is that technological modernization often seriously affects the traditional sources of meaning, often with no guidance on how to develop a new set of symbols to provide new meanings, for large numbers of people (and not always in the Third World).

Ellul on Technicization and the Possibility of Religious Resistance

Jacques Ellul is also widely recognized for his strictures on the alleged negative influence of technology on culture. Ellul's emphasis is not on culture generally but on religion as an antidote to the "technique" he sees as crowding out spontaneity and value in all realms.[25]

Ellul is almost always referred to in American intellectual circles as espousing a single idea—that of autonomous technology. Interpreters sympathetic to Ellul say this is a distortion; some even go so far as to say that Ellul's seeming pessimism is really a warning. The situation he describes will inevitably get worse (they say Ellul means to say) *unless we act quickly and decisively*.

One recent interpreter of Ellul who says this is D. J. Wennemann:[26]

> It is my view that Ellul's thought has been misunderstood because of a lack of appreciation of the dialectical structure of his thought. . . . Reading *The Technological Society* in isolation, one can only conclude that Ellul is a technological pessimist of the most extreme sort. But this study must be balanced by contrasting it with its theological counterpart. Ellul has said that his plan of research was to oppose his study of technique with *The Ethics of Freedom*.

According to Wennemann, Christian revelation provides a possibility of absolute freedom, and (he goes on):

> Ellul's intention is to attempt to make this freedom present to the technological world in which we live. In so doing, he hopes to introduce a breach in the technical system. It is Ellul's view that in this way alone [as Christians] are we able to live out our freedom in the deterministic technological world that we have created for ourselves.

Here we have an extreme claim for the cultural value of religion; and, furthermore, it is possible to read the claim as calling for a return to traditional religion. However, it is also possible to see religion and other cultural manifestations as not reactionary but forward-looking—as struggling to provide new meanings in troubled times.

A GENERIC RESPONSE TO THE PROBLEM

John Dewey, often interpreted as the great American advocate of social engineering, is sometimes parodied as an advocate of technocratic problem solving. A technocrat, however, Dewey is not; it is not even correct to call him a social engineer unless one is very careful about the meaning of that term.

I think the best account of Dewey's philosophy that has been put forward—perhaps the best until now, as I will be saying in a moment—is Ralph Sleeper's *The Necessity of Pragmatism: John Dewey's Conception of Philosophy*.[27] That remarkable account, which follows Dewey's philosophical development from its earliest beginnings to what Sleeper views as Dewey's "mature philosophy" represented most articulately and comprehensively in *Experience and Nature* and *Logic: The Theory of Inquiry*, begins with the claim that, for Dewey, philosophy is "a force for change," an instrument for transforming the culture in which we live. And Sleeper ends, in a chapter he says shows "the integrity of Dewey's work and some of its ramifications," with the claim that Dewey's philosophy is fundamentally meliorist. In an insightful and sharp contrast, Sleeper notes how:

> Although Wittgenstein and Heidegger share something of Dewey's concern for the release of philosophy from the constraints of tradition, they share little or nothing of Dewey's concern with the application of philosophy once released. They have none of Dewey's concern regarding the practice of philosophy in social and political criticism.[28]

Larry Hickman, in *John Dewey's Pragmatic Technology*[29] goes one step further. Hickman's thesis is that Dewey's philosophy is explicitly and consciously a meliorist critique of our technological culture. Perhaps exaggerating Dewey's occasional hyperbolic expressions, Hickman says that for Dewey philosophy *is* a technology—an instrumentality—for the transformation of our technological culture.

Hickman begins with this somewhat provocative statement:

> It is the central thesis of this book that inquiry within technological fields —among which [Dewey] included science as well as the fine and the vernacular arts—formed the basis of and provided the models for Dewey's larger project: his analysis and critique of the meanings of human experience. And it is no overstatement to say that his critique of technology was the warp on which the weft of that larger project was strung.[30]

In saying that a critique of technology was Dewey's main tool, Hickman is being only slightly less provocative than in his claim that Dewey's larger project was to restore meaning to a culture that had rendered not only science but also workaday skills and even the fine arts "technological" (or technical). But Hickman does not just mean to provoke; he believes his account is completely faithful not only to Dewey's intentions but to his actual words.

Hickman makes a key contribution in his reinterpretation of Dewey in chapter three, "Productive Skills in the Arts." There Hickman takes up Dewey's theory of aesthetics, which makes no sharp distinction between fine arts and practical, including industrial, arts. While this chapter is generally faithful to Dewey's words, I think it misses some of the fire of Thomas Alexander's treatment of Dewey's aesthetics, as providing meaning for a culture, in *John Dewey's Theory of Art, Experience and Nature*.[31]

Alexander rejects interpretations of Dewey that focus exclusively or primarily on his instrumentalism. He quotes Dewey as the authoritative interpreter of his own thought:

> To the aesthetic experience . . . the philosopher must go to understand what experience is. For this reason, . . . the theory of esthetics put forward by a philosopher . . . is a test of the capacity of the system he puts forth to grasp the nature of experience itself.[32]

Dewey and Alexander are here using both terms, "aesthetics" and "experience," in broad senses. Experience means the broad social experience of a community—what others would refer to as "popular culture" as well as "high culture"—and aesthetics includes not only the fine arts but also progressive religious attitudes and anything that creates meaning for a community:

> To keep experience from being treated always as a form of cognition . . . , one needs to articulate a position where the larger issues of human meaning and value contextualize the pursuit of knowledge. Knowledge is only possible because we can respond to the world as a dramatically enacted project in which meanings and values can be won, lost, and shared.[33]

Here is Alexander's summary of Dewey's views on the broad meaning of culture as *religious*:

> This aspect is precisely what Dewey describes in *A Common Faith* as the religious *quality* which experience may have. Dewey thereby is able to distinguish "the religious," which is a non-cognitive quality, from the particular doctrinal beliefs which constitute religions.[34]

For Dewey, "the religious" is the meaning-creating accompaniment, indeed the goal, of social activism. According to Alexander—and I think he is unquestionably correct—the progressive social activism that Dewey espouses in the name of "scientific method" or "instrumentalism" can only be progressive if it aims at creating meaning for a particular community struggling to achieve meaning in troubled or problematic times.

If we recall Hickman's basic thesis, we should recognize immediately that Hickman would be sympathetic with Alexander's account—indeed, that he would push Alexander to see that, for Dewey, aesthetics properly understood should provide the basis or motivation for a critique not just of any culture but specifically of our technological culture.

Chapter seven brings Hickman's book to a conclusion by examining social and political ramifications of Dewey's critique of technology and technological culture. This discussion of Dewey's political views has a certain intellectualist flavor to it that does not exactly square with Dewey's lifelong activism.

Though Hickman seems to have captured the essence of Dewey's philosophy remarkably well (as had Sleeper before him), I think it is Cornel West, in *The American Evasion of Philosophy: A Genealogy of Pragmatism*,[35] who best captures what the activism of Dewey's philosophy would require today. West claims to be improving upon, even going beyond Dewey to a "prophetic pragmatism" that learns more from Marxism than Dewey was willing to do, but I think West is more correct when he also claims that his view is a culmination of a pragmatic tradition in American thought that has Dewey as its pivotal figure but reaches its culmination in the intellectual and cultural environment of the United States today—and that for historical reasons.

The particular cultural context of our U.S. struggle for meaning in a technological world today that West concentrates on is a conflict in the universities over including in the canon of meaningful discourse new voices arising from black experience, women's experience, Hispanic experience—in short, of so-called "minority" experience generally. It happens that this struggle for the inclusion of multicultural diversity coincides precisely with the movement called "postmodernism"—a loose term found in cultural studies of all sorts, from critiques of art and architecture to philosophy, but also a term that is often associated with the jaded neo-conservatism of authors such as Richard Rorty[36] (West's principal target in his book). What West is saying we in the United States need today is a repudiation of this postmodern pseudo-culture, and we need to fight against it, faithful to Dewey's impulse, in a power struggle to broaden democracy and include the aspirations of those peoples West lumps under Frantz Fanon's label, "the wretched of the earth," both in U.S. culture itself and in the Third World.

People speak of ours as a technological culture; Bell and Marcuse and Ellul, as we have seen, would say it is a technological anti-culture. Dewey, as I interpret him, might well admit that technology today threatens traditional culture, but he would take this not as grounds for pessimism but as a challenge to social activism; as an invitation to create new symbols and new meanings for our troubled and problem-ridden world.

It is at this point that we need to raise a serious question. Are there activists around in sufficient numbers to carry on the struggle that West calls for? Is anyone *really* fighting to establish new meanings for our technological world? West himself believes that there are, and he points to the feminists, blacks, Hispanics, and other minorities who have been leading the battle for multicultural diversity in American universities. But what he really means to say is that the constituencies for which these academics speak—the real wretched of the earth—need to struggle with the cultural powers that be in order to create a new and more meaningful culture in a world threatened on every side by forces labeled "technological progress" that (in his view) are in fact rarely progressive. And when stated in such polarized terms, it is not clear that technological nihilism will not triumph over a new, progressive, more open technological democracy aimed at including the poor and the oppressed. This, however, would not have dismayed Dewey; a meaningful existence is not something to be taken for granted but something that must be won through arduous social struggle.

A PROGRESSIVE INTERPRETATION OF AMERICAN DEMOCRACY

As mentioned, my main project here is to argue that scientists, engineers, and others with technical training have social responsibilities. To make that argument, I depend on a particular conception of social responsibilities implicit in a democratic society, and that conception is based on a particular interpretation of the principles of the Declaration of Independence, the U.S. Constitution, the Bill of Rights, and later amendments that favor extending constitutional liberties to ever broader segments of society.[37] The connection with scientific and technical responsibilities rests on a denial of a standard distinction—between scientists' responsibilities *as* scientists and *as* citizens.[38] I believe scientists have social responsibilities as scientists and the same holds for engineers, physicians, etc.; and these responsibilities are related to their being citizens in a democratic society.

Further, I believe that, both in democracy generally and in technical communities, a small minority of public interest activists lead the way in holding up these responsibilities as banners for others to follow.

People sometimes speak as if there is a single interpretation of American constitutional principles. I do not believe that is so now, nor has it ever been so.[39] Among conflicting views of our constitutional ideals, I find most appealing the progressive view of the Constitution as a living document ever subject to reinterpretation as times change. Though strict-construction conservatives act as though this is not so, a progressive interpretation of the Constitution has been one of the competing claimants from the very beginning. This progressive interpretation (I agree with others) rests on five principal points.[40] (1) Admirable as was the original Constitution, it was also flawed in significant ways, especially in terms of the range of citizens granted rights. (2) The Bill of Rights, almost immediately, remedied some of the defects, particularly in establishing protections for the "natural rights" declared in the Declaration of Independence and presupposed in the Constitution. (3) Supreme Court decisions and the Civil War paved the way for the next set of extensions of civil liberties in the Thirteenth (1865), Fourteenth (1868), and Fifteenth (1870) Amendments, extending the rights and protections of citizenship to blacks (and, in the process, to all of us). (4) Twentieth-century amendments and Supreme Court decisions have further broadened the scope of these civil liberties, extending the right to vote to women and protecting all sorts of minorities against discrimination or infringements on freedom of choice. But (5) the struggle for greater and greater equality goes on.

The last point includes the need to make the rule of law more effective, to make majority rule more effective (e.g., by curtailing the power of money in elections), to realize still more fully the democratic ideal of equality (e.g., by providing forums for aggrieved minorities to dissent from majority rule), and to implement more fully the natural rights of citizens not only to equality of opportunity but to that minimum of so-called "welfare rights" that would allow citizens to exercise their rights.

This progressive interpretation of the Constitution can be applied explicitly to the tradition of citizen activism and dissent that, though a feature of American political life from the beginning, has flowered in the United States since the beginning of the civil rights movement in the 1950s and '60s. John W. Gardner, an outspoken advocate of citizen activism, notes how, "When this nation began in the 1770s, it had a population of about 3 million, yet it produced at least a dozen statesmen of extraordinary quality."[41] Gardner also quotes approvingly the famous statement about activism made by Alexis de Tocqueville in the 1830s:

> These Americans are the most peculiar people in the world. . . . In a local community in their country a citizen may conceive of some need which is not being met. What does he do? He goes across the street and discusses it with his neighbor. Then what happens? A committee comes into existence. . . . All of this is done without reference to any bureaucrat. All of this is done by private citizens on their own initiative.[42]

Gardner's point in citing de Tocqueville, however, was to note how, for a time in the first half of this century, Americans seemed to have lost their confidence in this approach. But:

> Then in the 1960s a feeling for citizen action reappeared with extraordinary vigor. It was foreshadowed in the 1950s in the civil rights movement. In the 1960s the students raised the cry of "participatory democracy." Among the poor the phrase was "community action." The peace movement, the conservation movement, the family planning movement emerged as potent elements in our national life.[43]

Public interest activism is certainly alive today. What I want to do is (a) point out how common it is in the scientific and technical community, and (b) emphasize opportunities in order to get more technical professionals involved.

ACTIVISM, SOCIAL RESPONSIBILITY, AND THE SCIENTIFIC COMMUNITY

A claim that scientists have special social responsibilities first reached a significant audience in North America in the aftermath of the United States' dropping of atomic bombs on Hiroshima and Nagasaki in Japan at the end of World War II. This response was institutionalized in *The Bulletin of the Atomic Scientists* and the Pugwash movement,[44] as just two examples. The call for social responsibility on the part of scientists, engineers, and biomedical researchers, among others, has been made repeatedly over the decades since.

There is, however, ambiguity and complexity in the call for scientific and technological responsibility—as shows up in Rosemary Chalk's *Science, Technology, and Society*,[45] a collection of papers from *Science* magazine, 1949-1988. Joseph Turner, in a 1960 editorial entitled "Between Two Extremes," argues for a middle position between an extreme that exempts scientists from any responsibility and an opposite extreme which demands that they foresee the consequences of and take responsibility for every piece of research before undertaking it. "This third position . . . seems to express the convictions of most of the persons in this country who are presently concerned with these problems . . . [; it] seeks the mean between the scientist's assuming too little responsibility for the consequences of his research and his assuming too much responsibility."

However complicated the notion, all the editorial writers and authors that Chalk includes admit that there is some social responsibility on the part of scientists. Bertrand Russell writes that, as scientists, "We have it in our power to make a good world"; C. P. Snow writes about "the moral un-neutrality of science"; Dael Wolfle ponders "a perplexing consideration of whether or not the scientist can separate his roles" as scientist and as citizen; finally, John Edsall and Anna Harrison focus on the primary responsibility of "getting at the truth" or maintaining "the integrity of knowledge"—presumably, either because that is what society expects of them or because the community of scientists expect it of themselves.[46]

However defined, by 1975, the Committee on Scientific Freedom and Responsibility of the American Association for the Advancement of Science had made it official in a report, *Scientific Freedom and Responsibility*,[47] that scientists and engineers have special social responsibilities.

In 1980, AAAS published a set of proceedings, *AAAS Professional Ethics Project: Professional Ethics Activities in the Scientific and Engineering Societies*,[48] that begins: "This final report of the AAAS Professional Ethics Project, prepared by the office of the AAAS Committee on Scientific Freedom and Responsibility, builds upon a long-standing concern within the Association about the importance of ethical issues in the development and use of science and technology." This focus on ethics had become the accepted version of the official line, and one of the major findings of the project is that more and more technical professional societies are including in their codes of ethics a directive to make safeguarding the public a matter of paramount importance.

In 1988, AAAS started a newsletter, *Professional Ethics Report*, that is the current voice of this official position.

These efforts are surely paralleled by similar institutional responses elsewhere. I know, for example, of similar efforts in Canada and Germany, and I suspect that almost any country that has a large-scale commitment to the promotion of science and technology will also feel the pressure of an alert citizenry ready to call the technical community to task when irresponsible behavior is alleged in connection with scientific research or engineering development.

I doubt the official statements mentioned above would have been forthcoming if it had not been for activists in the professional societies in the 1970s—typically under banners such as "engineers for social responsibility." I believe that these progressive scientists and engineers were strongly influenced not only by a progressive interpretation of Constitutional rights—especially the ideal of equality—but also by the post-civil-rights activists to whom Gardner refers.

I know many progressive scientific and engineering activists personally, and all of those I know are convinced that all scientists, engineers, physicians, etc., have social responsibilities. However, they also recognize themselves as a minority in speaking up publicly for their recognition.

A brief historical note: William Gilman's *Science: U.S.A.* faces an important issue squarely: "Who is to choose among the many [competing and expensive] science projects?" His answer is straightforward: *the people*, acting principally through their representatives in Congress, aided by scientifically and technologically expert staff. Gilman traces his source for this view back to this:

> Thomas Jefferson's answer to the would-be-elite of his day was that of a wise, worldly man, "And true it is that the people, especially when moderately instructed, are the only safe, because the only honest, depositories of public rights."[49]

I agree with Gilman echoing Jefferson. Today, progressive scientists, engineers, and other activists from the technical community should recognize that their work is supported by the community. Therefore, as a matter of course, the public has a legitimate say in requiring accountability on the part of the technical professions and other scientific institutions. Social responsibility is shown by the scientific and technological communities when they do not hide behind the scientist-as-scientist slogan, behind a mask of value-neutrality in the name of scientific objectivity. It is shown, instead, when they work with government regulators for the public interest, or when they get involved in public interest activism where government or private-sector technological ventures threaten to harm the public.

THE SET OF CONCRETE PROBLEMS ADDRESSED HERE

Few thinking people deny that there are evils in our world today. In this book, I am concerned primarily with a particular set of problems —those that have concerned antitechnology critics such as Ellul and Marcuse. (A preliminary outline can be found in chapter 2, but see also the table of contents, parts II and III.)

Any list of antitechnology authors might well begin with Kurt Vonnegut, Jr., in *Player Piano* or *Cat's Cradle*. This sort of antitechnology literature goes back at least to Mary Shelley's *Frankenstein* (1818), but it is perhaps most clearly epitomized in the story of "Hal," the computer who(?) takes command in Arthur C. Clarke's *2001: A Space Odyssey*. I am more concerned, however, with scholarly antitechnologists. These include not only Ellul[50] and Marcuse,[51] but also the obscure but deep philosopher Martin Heidegger,[52] Heidegger's American disciple, Albert Borgmann,[53] and techno-political scientist Langdon Winner.[54]

On the other side, defenders of science and technology have responded to these attacks. The best example is Samuel Florman.[55] John Passmore, more balanced than Florman, has written a passionate defense of rationality, claiming that the antitechnologists favor irrationality.[56] Emmanuel Mesthene maintains that, on balance, technology has brought with it more opportunities than evils.[57] There is even an explicit philosophical attack on "no-growth futurism," by Edward Walter.[58] Daniel Bell, with whom I began, is (possibly erroneously) taken to be the leading defender of the ascendancy of technical elites.[59] It should be noted that none of these defenders of technology goes so far as to say that there are not serious technosocial problems.

Nor does the matter stop with critics and defenders of technology. Marxists, as pro-technology as Florman, have argued that only under socialism can technology be effective in countering current social evils.[60] This used to be the official line in East Bloc countries, with their so-called "scientific-technological revolution" theories.[61] As these countries undergo dramatic reforms, there seems to be no lessening of their commitment to technology or a technology-based economy.

At the opposite end of the spectrum, many conservatives are opposed to the regulation of technology, which would seem to make them at least implicitly pro-technology.[62] Many are explicitly pro-technology in the sense that they support continued research on the latest high-technology weapons and intelligence systems despite the end of the Cold War.

Scholarly conservatives—assuming that evil is part of the human condition—are the only intellectuals I know of who tend to play down technosocial problems; according to them, social problems are not alleviated but exacerbated by social engineering.[63]

To summarize: what this book is about is technosocial problems and how activists—including scientists and engineers and others in the technical community—can be motivated to do something about these problems. I argue that scientific and technical activists, crusading under the banner of social responsibility and working alongside non-scientific public interest activists to make ours a better world, provide us with some hope. In any case, this approach seems to me to have more to offer than the gloomy prophecies of the radical critics of technology, and also more to offer than the more-of-the-same proposals of people defending the technological status quo.

2

Ethics as Social Problem Solving

Do engineers have social responsibilities? Whence do they arise? What do they entail? . . . The answer (albeit not conclusive) that emerges from [my applied-ethics] analysis is that engineers do not have special social responsibilities as engineers, but they do have responsibilities as persons.

—Deborah Johnson

The order of the universe that we live in is *the moral order. It has become the moral order by becoming the self-conscious method of the members of a human society. . . . The world that comes to us from the past possesses and controls us. We possess and control the world that we discover and invent. . . . It is a splendid adventure if we can rise to it.*

—George Herbert Mead

I started my philosophical career as a Thomist—though I always thought of St. Thomas himself as principally an interpreter of Aristotle. As a student of Thomistic philosophy, I grew up on slogans: "Philosophy bakes no bread." "It is useless—completely non-utilitarian." "Philosophy is like pure science at its best, the pursuit of truth for its own sake, closer to the liberal arts than to engineering." Imagine my surprise, then, when I first read—then studied, and finally became enamored of—the activist philosophers George Herbert Mead and John Dewey. This chapter sets up a contrast between the sort of activist philosophy espoused by Mead[1] and a common distinction between technical and civic responsibilities.[2]

I discovered Mead first, and what impressed me most about him was the way his philosophy and his life meshed so perfectly. According to David Miller in his intellectual biography, "Mead was continually concerned with civic affairs"[3]—with the schools, with labor negotiations, with settlement houses—and he held offices in several civic organizations. He was also a lover of the arts, and he and his wife entertained in their home all sorts of social activists, including leaders in the women's suffrage movement.

John Dewey, if anything, was even more of an activist. As Alex Michalos notes: "Perhaps the easiest way to reveal what Dewey stood for in his long life . . . is to simply list the organizations to which he belonged."[4] Michalos is here reviewing Gary Bullert's *The Politics of John Dewey*,[5] and he goes on to cite at least twenty organizations of major importance that Dewey founded, headed, or was associated with all of this while doing enough academic and public philosophy publishing to fill over thirty volumes in a new critical edition.[6]

I reduced my "cognitive dissonance," to borrow a term from psychology, and consoled myself by recalling a remark from John Herman Randall's *Aristotle*: "It is certain that were Aristotle to write today an *Ethics* and a *Politics* for contemporary Americans, he would not elevate knowing above practical action."[7]

What I do in this chapter is try to persuade at least a few of my fellow philosophers of technology that *any really worthwhile ethics of technology must have a practical payoff*. (I take up in chapter 16 my relationship to mainstream academic philosophy.) I can express the same goal in a different way. Willem Vanderburg[8] cites as the basic slogan of Jacques Ellul, "Think globally; act locally," and most people would say the philosophical part is the global thinking. I think, rather, that the philosopher must both think and act *locally*. I defend this view here, then show in chapter 3 how it might work out in practice.

PROBLEMS OF TECHNOLOGICAL SOCIETY

I begin my defense of my mildly paradoxical thesis with a second autobiographical note. What motivated me to turn from interests in abstract philosophy of science to concrete philosophy of technology was social problems: the militarization of science, claims of chemical companies to produce "better things for better living through chemistry" while they were swamping us with toxic wastes, and so on. I assume that many others who turned from academic concerns to philosophy of technology were similarly motivated.

What I propose is a thought experiment. Assume that, say, half of the most prominent technological problems (including thousands of local manifestations) have been at least temporarily dealt with in at least a somewhat successful way. What would be the reactions of those philosophers who have attempted to provide an ethics of technology?

As a starting point for discussion, I propose this list of ten representative sets of contemporary social problems that are arguably associated with technology.

A. Issues of survival:
1. Nuclear weapons;
2. Toxic wastes and other major ecological threats;
3. Genetics and computer threats to what it means to be human.

B. Issues related to the "good life":
4. Technoeconomic injustices, including growing disparities between rich and poor classes and rich and poor nations;
5. Hazards associated with high-technology work, as well as

boredom and lack of fulfillment in those jobs;
6. Problems of alienation and associated problems of families in high-technology society, especially in urban areas;
7. Declining quality in schools, in terms of both technological illiteracy and lack of civic preparedness;
8. The contemporary health care crisis;
9. Political alienation in an era of technological domination of elections; and
10. Commercialization of the arts and of traditional "high culture."

PROPOSED ETHICAL SOLUTIONS

Quite a few philosophers of technology have proposed ethical answers for technosocial problems. I would list some of the main ones as follows:

1. For all technosocial problems the standard answer in the scientific-technological community is either engineering ethics or technology assessment—the latter usually dependent on some version of risk/cost/benefit analysis. Kristin Shrader-Frechette believes that such technical exercises can be transformed into an ethical response[9] with the appropriate addition of an equity dimension following John Rawls.[10] (I spell out my relationship to academic ethics in chapter 16.)

2. At the opposite extreme is Hans Jonas, who has developed a sort of neo-Kantian categorical imperative of fear or caution for the technological age.[11]

3. I do not consider myself competent to say what Martin Heidegger's ethical response to the technological age would be—assuming he had one.[12] However, one Heideggerian, Daniel Dahlstrom, thinks he can. Recalling Heidegger's early claims that he was not doing ethics—that, instead, he was trying to comprehend the *presuppositions* of any ethical or humanistic response to the modern age—Dahlstrom nonetheless attempts to formulate a Heideggerian technological ethics. As an example, Dahlstrom uses the "technology" of eating. In our day, he says, "The business of displaying grocery products in supermarkets and the marketing of menus in the mass media seem to force the presence of food upon us, yet the act of being food, its fragile presence and absence is obscured." Dahlstrom says this against the background of the technological problems of "junk food" and "fast food," or of farm surpluses being destroyed while Africans are dying of starvation. The ethical response Heidegger would make, Dahlstrom says, is simply to recognize this: "The essence of modern technologies of food and consumption . . . constitutes a way being unfolds and discloses itself, that is to say, a way of nourishing and surviving, however much it also obscures this fundamental event." What is this event viewed ethically? Dahlstrom concludes: "[It] is a ritual celebration (eating) of a gift (food) of the earth and the sun to the human species."[13]

4. Neo-Heideggerian or post-Heideggerian philosopher Albert Borgmann[14] goes even further than Dahlstrom. He is convinced that he has captured the phenomenological essence or spirit of the technological age in what he calls the "device paradigm." People in the modern era have tended, with increasing success as devices have proliferated, to try to "disburden" themselves of the difficulties of life, to assure "the good life" by means of ever-increasing consumption of commodities. Borgmann thinks that the only ethical solution for technology's problems is public (his term is "deictic") discourse in which defenders of "focal things" will manage to balance the stridency of the appeal of devices without destroying the advantages that particular technologies can provide—even for "focal" activities. Borgmann is thinking of things like Dahlstrom's ritual celebration of eating; however, he cites many other examples which circulate in literatures of isolated subcultures within technological society while currently leaving relatively unaffected its consumption orientation. (I reply in some detail to Borgmann and other neo-Heideggerians in chapter 12.)

5. If I read Jacques Ellul correctly, he is convinced that no ethical response can be effective in the face of autonomous technique or the "technical ensemble."[15] (I treat Ellul's views indirectly in chapter 12 as well.)

6. Philosophers I would refer to as Ellulians, such as Gilbert Hottois, believe that under certain conditions an effective ethical response to "technoscience" (if not to particular technological problems) is possible. Hottois says:

> What we are dealing with in the problem of the relationship between technoscience and ethics cannot possibly be captured in a supposed opposition between two ethical attitudes, two moral conceptions, or two value systems. The stakes are much more profound. . . . Any counterweight to the nihilistic power of technoscience cannot come from one or another moral system but only from a new burst of moral consciousness in general, from the originating source of every possible moral system. That is how deep the level of the encounter is between ethics and technology.

Hottois then adds his main point:

> An effective core of a renewal of ethical consciousness awakened by technoscience can be nothing other than the renunciation of power. . . . The new ethical consciousness must be that of non-power [as Ellul notes]. And here we are at the heart of ethics itself, not of some particular expression of it conditioned by a need to respond to the technical context. Indeed, what is "being moral" if not the ability to decide freely not to do everything one is capable of doing, . . . freely to put limits on freedom?[16]

7. Another follower of Ellul, Daniel Cérézuelle, says much the same thing in the more limited context of bioethics as an inadequate response to the problems associated with high-technology medicine. Cérézuelle

begins with a typically Ellulian premise: "An approach to biomedical technology which is limited to codifying particular [ab]uses is insufficient[;] . . . it is necessary to engage the technical processes in their ensemble." Cérézuelle puts this strain of Ellul's thought in the context of "socialist reformers of the last century," who "were correct in thinking that the development of industry, which was completely transforming society, could not be controlled by limiting themselves to the legal codification of relations between individuals." Cérézuelle's version of the non-power conclusion is this: "Taking into account the irremediable character of human and social imperfection, it seems reasonable to limit knowledge and technology, at least provisionally, this side of that threshold beyond which we no longer control the effects." This is despite, or perhaps precisely because (as Cérézuelle goes on to say), such a moratorium "is naturally repugnant to a culture which sees in the indefinite and unlimited pursuit of science and technology an expression of human greatness which nothing should impede."[17]

8. Cérézuelle refers to Ellul's acknowledged Marxist background. Can Marxism be viewed as an ethical response to problems of technological society? (I take up this issue in chapter 11, where I ignore the vicissitudes of Marxism after the so-called end of the Cold War.)

In a study devoted to *Marxism and Ethics*, Eugene Kamenka concludes: "In the field of ethics, Marx himself may be regarded as a social critic rather than as a moral philosopher"—though Kamenka grudgingly admits "that a sensitivity to history and to social questions, such as Marx had, is invaluable to the moral philosopher."[18] Whether or not it is "ethical" in some academic or analytical sense, however, Marxist Socialism can be presented as the only effective way of dealing with the problems of technology. For instance:

> The effects of technological development . . . are not the same in capitalist and socialist society. In socialism, the notion of "technical advance" is connected with the idea of [revolutionary] social progress. . . . This does not mean that problems do not exist in this society; it is only that they are no longer fatal. There are favorable conditions for dealing with them successfully.[19]

This certainly expresses the old official line of the Communist Bloc countries, but die-hard Marxists continue to maintain the view—at least as something to be realized at some unspecified future time.[20]

9. The ethics-of-technology views listed so far tend to be rather sweeping in their scope. One Neo-Marxist, David Noble, is both less sweeping and much less sanguine about the prospects for liberation under technological socialism. His remedy for technological problems under any political system—he calls his approach "present tense technology"—is for the workers to seize control of their workplaces wherever they can in a self-conscious neo-Luddism. According to

Noble, this is not so much an ethical approach as it is a matter of recognizing the political nature of the problems: "The real challenge posed by the current technological assault is for us to become able to put technology not simply in perspective but aside, to make way for politics. The goal must not be a human-centered technology but a human-centered society."[21]

Two more views, which share Noble's narrower scope, explicitly oppose the sort of progressive activism I am defending in this book.

10. Langdon Winner is explicit in disparaging this activist approach: "One can assume the role of a firefighter, select one or more of the areas of life in which technological change looms as a problem, and set one's goals to improve things in that sphere. . . . But a therapy that treats only the symptoms leaves the roots of the problem untouched."[22] (As I will note when I discuss his views in detail in chapter 12, it is unclear to me how Winner's "epistemological Luddism" of "dismantling particular technologies" as necessary and as possible amounts to anything different from social and political activism.)

11. Steven Goldman, taking aim at an earlier version of my Mead-Dewey public interest activism, describes it as refusing to address "the underlying causes of technology-related social problems in favor of ameliorating the effects." And Goldman says the approach must confront "an interminable series of challenges," unless it aims at the "common underlying cause [of the problems] instead of aiming at what is unique to [the] respective surface problems."[23]

WHY I THINK THE MEAD-DEWEY APPROACH IS THE BEST WE HAVE

In chapter 1, I referred to Ralph Sleeper's excellent interpretation of Dewey's philosophy as fundamentally meliorist. There I quoted Sleeper's contrast of Dewey with Heidegger and Ludwig Wittgenstein. According to Sleeper, Heidegger and Wittgenstein "have none of Dewey's concern regarding the practice of philosophy in social and political criticism."[24] Earlier in his book, Sleeper had noted how this "accounts for his [Dewey's] . . . pervasive sense of social hope. It accounts for . . . his dedication to the instruments of democratic reform; his historicism and his commitment to education; his theological agnosticism and his lifelong struggle to affirm the 'religious' qualities of everyday life." I suspect it is clear to anyone who has read Dewey carefully that the sorts of problems Dewey wanted to attack with his transformed, meliorist philosophy were precisely the sort I listed earlier.

Mead did not live nearly as long as Dewey, and the social problems to which he addressed his equally meliorist philosophy were those of the first three decades of the twentieth century, prior to the high-technology period of "post-industrialism" or the "scientific-technological revolution." But the spirit of his philosophy is the same. And, as seems to be the case more often than not, Mead is much clearer than Dewey in stating the theoretical underpinnings of their shared approach.

According to Mead:

> The order of the universe that we live in *is* the moral order. It has become the moral order by becoming the self-conscious method of the members of a human society. . . . The world that comes to us from the past possesses and controls us. We possess and control the world that we discover and invent. . . . It is a splendid adventure if we can rise to it.[25]

In other words, society acting to solve its problems in a creative fashion is by definition ethical.

Traditional definitions of ethics are inadequate, Mead thinks, and he grounds his social-action approach on this inadequacy. This is emphasized by Hans Joas, in a recent reinterpretation of Mead: "Mead and Dewey developed the premises of their own ethics through criticism of utilitarian and Kantian ethics." Specifically, "In Mead's opinion, the deficiencies of utilitarian and Kantian ethics turn out to be complementary: 'The Utilitarian cannot make morality connect with the motive, and Kant cannot connect morality with the end.'"[26] Utilitarians, who base their view on people's self-interest, according to Mead, fail to provide an adequate grounding for altruistic social action. Kant, on the other hand, fails to see that the right way to do one's duty is not predetermined; it must be worked out in a social dialogue or struggle of competing values.

In both Dewey and Mead, ethics is simply the community attempting to solve its social problems in the most intelligent and creative way its members know how. In a technological world, ethics is community action attempting to solve urgent technosocial problems.

CONCLUSION

I conclude by repeating my earlier claim: as an ethical response to technology, I would be satisfied to have half of today's major sociotechnical problems solved, even if it appeared that society would not thereby be treating the underlying cause or causes of the problems. What would be the response of the ethics-of-technology advocates on my list?

i. Ellul and Hottois and Cérézuelle would probably still not be satisfied. They would say we must, as our ethical response, say "no" to technological development as a whole. However, they also want us to say "no" to particular technological developments that are especially threatening, and I find it difficult to see how this remedy means anything more than urging particular groups to say "no" to particular technological developments. Supposedly global thinking, then, ends up leading to local action.

ii. Heideggerians such as Dahlstrom and Borgmann offer very complex analyses of the double effect of modern technology (or of the "device paradigm"): that it both conceals and reveals the true nature of modern life. This would again seem to suggest that solving half of our problems would not be enough, that we need a wholesale transformation of modern society. But their ethical imperative turns out to be this: We should recognize and concentrate on "focal" things and processes such as "the culture of the table." We will thus see at once the inhumane character of "fast food" or of destroying surplus food while people starve.

While this seems to aim at wholesale revision of our culture, if it has any ethical meaning in concrete terms, this must be a call for particular groups of individuals—for example, Borgmann's "underground" groups of people already devoted to focal things and practices—(a) to come to such an understanding, and (b) to persuade others by way of public discussion (Borgmann's "deictic discourse"). Again, this seems to me to amount to little more than piecemeal social activism.

iii. Neo-Marxist Noble enjoins workers to follow the lead of erroneously-maligned Ned Ludd and take control of their would-be high-technology workplaces whenever and wherever they can. I am not sure whether Noble would follow the lead of earlier Marxists in saying that nothing short of a worldwide workers' revolution will get the job done. Assuming he would not, his neo-Luddism seems indistinguishable (except perhaps in its motivation) from piecemeal technosocial problem solving—something that can also be said about the proposals of Goldman as well as Winner.

iv. More traditional Marxists should refuse to play my game. They should deny categorically that any of the other technosocial problems can even be imagined to be solved without the elimination of capitalism or, more generally, the class struggle between the owners of the means of production and oppressed workers. They should, in other words, say that my thought experiment—assuming that half of our social problems have been solved—is utopian. (I take up this objection in chapter 11.) Still, if Marxists will grant my thought experiment *at least for the sake of argument*, then they should also admit that, under those circumstances, the claim that class divisions are the only root of technosocial problems has a hollow ring. And, in the meantime, working on social problems that are easily separable from the class struggle should make sense.

v. This leaves us, as an ethical response to technology, with only two of our original candidates: either a sanitized version of risk/cost/benefit technology assessments (*à la* Shrader-Frechette) or with Jonas's neo-Kantian categorical imperative for the technological age. Many other neo-Kantian and utilitarian responses to specific technological problems have been put forward recently in the field called applied ethics. (This is the context of Deborah Johnson's quote used as a headnote. I provide my response to academic applied ethics in chapter 16.)

We are, in the end, not limited to my original list of ethical responses to the problems of technological society. We can, without apology, espouse the piecemeal, public-interest-activism approach modeled after the social ethics of Mead and Dewey. That may not satisfy many philosophers, but to me it sounds as though, with respect to ethics and technology, the situation is what Winston Churchill said it is with democracy. According to the often quoted saying, democracy is the worst form of government—except for all the others. Similarly, a piecemeal approach to social problem solving may seem the worst sort of ethics for our technological age—except for all the rest.

I do not, however, want to end on that note here. I believe one can make a positive defense of a social ethics of technology. What this means for me is that in fact there is some hope that some of the major social problems of our technological age are being solved. A recent study of reform politics and public interest activism says just that:

> Throughout the [United States], myriad progressive groups have been mobilizing and acting on behalf of crucial issues largely outside the glossy mainstream of media politics: the variety of church, campus, and community organizations mobilized around issues of U.S. policy in South Africa and Central America as well as nuclear arms policy; the increasingly effective women's and gay-rights movements; the growing numbers of radical ecologists and advocates of "Green Party" politics; the renewed efforts to mobilize blacks, ethnics, and the multitude of the poor by Rev. Jesse Jackson and others; the diverse experiments of working people both in and out of labor unions to reassert themselves; and the legions of intellectuals committed to progressive economic and social policy formulation—all have constituted elements of an increasingly dynamic movement to build an eclectic base of progressive politics in the nation.[27]

This quote from Michael McCann puts the case for progressive reform generally. In this book, I am concentrating on the kinds of reforms technically-trained professionals might be able to bring about. I take up specific examples in parts two and three, focusing on seven of the ten representative types of technosocial problems mentioned earlier. Part two addresses general problems, such as education, health care, and politics. Part three focuses on problems specifically related to technology: biotechnology, computers, nuclear weapons and nuclear power, and problems of the environment. In each case, I show how likely it is that no real reform will actually take place unless technical professionals are willing to go beyond what is demanded by their professions and get involved with activist groups seeking to bring about more fundamental change. (I will make the same claim, with respect to academic philosophers, in the final chapter of the book.)

Before launching into my demonstration of how this might work out in practice, I feel a need to provide a sample case. What I have chosen for this purpose is the case of professionals attempting to deal with problems of families in our technological world. Here we see clearly displayed the combined power (if they get involved in activist ways) and weakness (if they do not) of that set of professionals most people would see as most likely to get involved in activism in our culture. (I note, in passing, that some of them are less likely to get involved today than they might have been a decade or two ago.) What I hope to show by this means is a *pattern*: trained professionals—in this case, social workers and other "helping" professionals—who attempt to deal with the problems they are trained to address are helpless to get their professional goals accomplished if they do not go beyond *mere* professional work and become involved in activist coalitions with people outside their professions.

In parts two and three, then, I try to show this same pattern with respect to technical professionals.

To end this chapter, I will just summarize what I have said. There are a great many social problems in our technological world. Many ethical solutions have been proposed. But in the end none of them seems as likely to be a solution as the approach of G. H. Mead and John Dewey—namely, a piecemeal approach that would urge philosophers to work alongside other activists in actually dealing with the problems that face us. Other ethics-of-technology approaches might also work, but in my view that can only happen if their practitioners become as actively involved as Mead and Dewey were.

3

Social Workers: A Sample of What Is to Follow

We have gathered at the World Summit for Children to undertake a joint commitment and to make an urgent universal appeal—to give every child a better future. . . .

Each day countless children around the world are exposed to dangers that hamper their growth and development. They suffer immensely as casualties of war and violence; as victims of racial discrimination, apartheid, aggression, foreign occupation and annexation; as refugees and displaced children, forced to abandon their homes and their roots; as disabled; or as victims of neglect, cruelty and exploitation.

—*United Nations Declaration on Children (1990)*

Is not the rise of an urban civilization which accompanies the advance of industrial civilization a true challenge to the wisdom of man? . . . Behind the facades, much misery is hidden, unsuspected even by the closest neighbors; other forms of misery spread where human dignity founders: delinquency, criminality, abuse of drugs and [commercialized] eroticism.

—*Pope Paul VI*

Representing a traditionalist view of the family, Pope Paul[1] here is lamenting the horrors of urban slums amidst technological affluence. The United Nations Declaration on Children[2] narrows the focus of the same issues to problems of children. Pope Paul goes on to lament the generation gap between children and parents, unemployment, and rootlessness—along with what he sees as the excesses of feminism.

From the opposite end of the moral and political spectrum, Letty Cottin Pogrebin finds a different set of evils affecting families—and especially children—today. Pogrebin thinks the worst enemies of a healthy family life are reactionary defenders of the traditional patriarchal family—including the popes. She summarizes contemporary threats to the family in America under the heading, "Pedophobia: Ambivalence and Hostility toward Children":

> In our own era [as opposed to the days of routine forced child labor and religion-sanctioned beatings], pedophobia is more subtle. It is also more pernicious because it exists within the lie that Americans are a child-loving people. Today, we protest the cost of public education, child health programs, or food stamps. We object to children's shelters being located in our neighborhoods, where they might lower our property values. Public accommodations ignore the existence of children and their needs.[3]

In Pogrebin's view, the grossest evidence is to be found in widespread child abuse, "the physical and sexual violations committed in families, on the streets, and in child pornography." She recalls the terrible statistic that ten percent of all American children are *reported* as abused—adding that "authorities say child abuse is grossly *under*reported." Pogrebin finds evidence of institutionalized hatred of children in widespread toleration of truancy and undereducation—especially among children of immigrant workers—as well as in the Reagan Administration's condemning of more and more children to poverty by way of cutbacks on programs to fight malnourishment or to provide health and social services for the poor.

Between these complaints from the right and the left, even moderate social critics point out how, in our technological world—where so many women must work outside the home or are raising children on their own—a majority of children are left to fend for themselves at home alone after school.[4] Whether or not, at least at times, this constitutes neglect, no one seriously denies that there *is* widespread neglect and abuse, including sexual abuse of children.

In this chapter, I consider what is being done to deal with this and similar issues. My focus in on people trained to deal with these issues, principally social workers. In the end, what I conclude is that professional social work alone, no matter how good the agency, is never enough. Professional efforts for the poor and dispossessed, for those who cannot measure up to the demands of a high-technology world, for those who are abused or neglected, are never enough. Where any lasting reforms are accomplished, it is because of broader social movements, typically instigated by social activists. Social workers can and often do join these movements. But the point is that they cannot be effective in their professional work in any other way.

I take this almost obvious fact as a pattern for what is needed with respect to the problems of technological society addressed in parts two and three. Here it is most obvious that professional work alone is not enough, that it must be supplemented by vigorous activism. In the cases to be discussed later, this will not be so obvious. But I will maintain, in each case, that for the solution of any technosocial problem, professional activity alone is not enough; activism is needed. The present chapter, then, completes the introductory part of the book with a sample of the pattern to be followed later.

PROBLEMS ADDRESSED

I have selected three sets of problems for consideration in this chapter: neglect and abuse of children, alarming rates of teenage pregnancies, and equally alarming unemployment rates among young African-American males too often unprepared for the jobs that the economy has to offer today. The problems of families today, in our society, are legion. One must choose, and I think the issues chosen are representative—and at least the first two represent examples of deep involvement by social workers, child advocates, and their agencies.

CHILD ABUSE AND NEGLECT

According to the most authoritative sources, there are almost a million substantiated cases of abuse and neglect of children in the United States each year.[5] This is not a new problem. Children have been grossly mistreated throughout history.[6] Some might naively hope that our supposedly enlightened age would have a better record, but that does not seem to be the case. Perhaps we are no worse than earlier ages, but we do not seem to be much better either.

If we turn from abuse and neglect in general to child sexual abuse, the picture is just as bleak. The most cautious and careful of the estimators of the incidence of child sexual abuse, David Finkelhor, summarizes the data for us. The lowest reliable estimates for the United States say that about fifteen to twenty percent of girls and five to ten percent of boys are sexually victimized each year;[7] the highest estimates, with a limited sample but based on in-depth interviews, raise the estimates for sexual abuse of girls to almost forty percent (over fifty percent if we count unwanted advances, excluding sexual contact).[8]

These seem to be alarming rates, even if we take into account American subcultures whose mores tolerate or even condone the beating of children as a part of "education." Presumably the reported cases of abuse take this factor into consideration, and in any case subcultural toleration of sexual abuse would be frowned upon by the vast majority of Americans and would be dealt with (where discovered) severely by the courts. Certainly these issues are taken seriously by social agencies, and it is to one of these that I now turn.

The Child Abuse and Prevention Act of 1974 provided for a National Center on Child Abuse and Neglect in the Department of Health and Human Service's Children's Bureau, and more specifically in the Administration for Children, Youth and Families, Office of Human Development Services. Its functions were to include generating new knowledge and improving existing programs; serving as an information resource; coordinating federal efforts; and assisting states and other jurisdictions in implementing child abuse programs.

The National Center on Child Abuse and Neglect (NCCAN) in its first ten years funded almost four hundred projects, ranging from research on causes and prevention to treatment of abused children and abusing parents. Its focus included sexual abuse along with other kinds of abuse and neglect. A series of national conferences were held. There is an Advisory Board on Child Abuse and Neglect, an interagency board, that coordinates all the federal efforts. And awards to states and other jurisdictions to establish or strengthen existing programs have regularly been in the $5 million-plus range each year.[9] In short, a sizable bureaucracy now exists, in Washington and in states, counties, and cities, to combat child abuse and neglect and to help both victims and perpetrators.

One thing that stands out in the literature on dealing with child abuse and neglect is that the professionals recognize that they cannot get the job done by themselves. One report asks: "Why should we implement prevention on a community level? Why can't this be done utilizing only trained specialists such as social workers, physicians and teachers?" And the answer is immediate and straightforward: "Professionals do not have the capacity to run prevention programs by themselves." What is needed, these authors (professionals themselves) say, is to "acquire broad community support and institutionalize our prevention efforts within major community systems." They go on: "All of the major community forces need to be utilized in this process, including service clubs, business and civic leaders, church groups, health professionals, trade unions, legislators, educators, . . . and parent groups"—all working alongside the child welfare professionals.[10] And even where something approximating this ideal is realized, the problems continue.

A critic could say that the reason all these efforts, by professionals and community groups supporting them, do not make a significant dent in the problem is that they are not getting at the roots of the problem; they are dealing only with superficial symptoms. In my opinion, this radical view should not be totally discounted. The community coalition building around issues of neglect and abuse may be too timidly avoiding stepping on the toes of institutions (for instance, some religious denominations) that may, at least indirectly, encourage child abuse even while they claim to be working against it. At any rate, if child abuse rates are going to be reduced significantly in this country, more vigorous activism is going to be needed than the coalition building we have seen so far.

PROBLEMS OF TEENAGE PREGNANCY

According to a National Academy of Science study, *Risking the Future* (1987),

> Regardless of one's political philosophy or moral perspective, the basic facts are disturbing: more than one million teenage girls in the United States

become pregnant each year, just over 400,000 obtain abortions, and nearly 470,000 give birth.[11]

This quotation is used as the starting point for a major, two-volume study of what can be done about this problem, published in cooperation with the National Organization on Adolescent Pregnancy and Parenting (NOAPP). The authors are Jeanne Warren Lindsay and Sharon Rodine.[12] NOAPP is not a bureaucracy but a national network devoted to supporting those social agencies, public or private, that are working to combat the problem of teenage pregnancy or to help teenage mothers (sometimes also fathers) before, during, and after the birth of their babies.

The sweep of the focus of the NOAPP study is apparent in the chapter headings of Book Two:

1. Understanding Adolescence
2. Prevention in the Family and the Community
3. Primary Prevention—In the Schools
4. Working with Sexually Active Teens
5. Considering Alternatives
6. Focusing on Males
7. Serving Pregnant Teens
8. Helping Teen Parents
9. Providing School Services
10. Caring for the Next Generation

Also included is an appendix detailing resource programs throughout the country, including offices to contact in all but one of the fifty states.

With this problem as with the problem of child abuse, there is a recognition of the need to go beyond professional social work and child welfare efforts. The first book, *Teen Pregnancy Challenge,* includes a chapter with the title, "Advocating to Policy Makers." The second book has a corresponding chapter on developing a community coalition. Under the advocacy heading, Lindsay and Rodine are very careful to point out what can and cannot be done, whether under Internal Revenue Service rules or under the specific guidelines of funding sources. Under the coalition heading, they stress the importance of coalitions "in building a climate that makes adolescent pregnancy prevention a priority," and they say such coalitions can help in public awareness, in coordinating resources, in training, in advising public officials, in monitoring services, and in drumming up support for the effort.[13]

This may seem slightly more activist, slightly less "establishment," than the activism with respect to child abuse, but it is still clearly in the mainstream. Lindsay and Rodine do talk about the causes of teenage pregnancy, but the principal ones they focus on are individual—from lack of self-esteem to teenage risk-taking, from peer pressure to ignorance or bad attitudes about birth control—rather than social. They do mention dysfunctional families as another cause but do not say anything about social conditions that lead to this dysfunctionality; and they talk about media influences—e.g., radio and TV, including suggestive lyrics and programs that suggest that "everybody's doing it"—but again without considering any of the deeper issues about the influence of the media on life in a technological culture. Surely advocacy and coalitions are needed if teenage pregnancy rates are going to be lowered, but equally surely some more vigorous activist measures are needed before the root causes can be addressed effectively.

UNEMPLOYMENT AMONG BLACK YOUTHS

The third problem area considered here—alarming unemployment rates among young black men—introduces two other catastrophic social evils of our supposedly enlightened age, high crime rates and racism. It is, of course, not the case that young black males alone are responsible for the high crime rates of recent years. But that is the common misperception, and that is what indirectly introduces the issue of racism.

There is an excellent study of the unemployment aspect of this complex issue: *The Black Youth Employment Crisis*,[14] by Richard Freeman and Harry Holzer. That mammoth study, funded by the Rockefeller and Ford Foundations, among others, begins with this premise:

> In recent years, with the onset of equal employment opportunity, affirmative action, and other government and private efforts to reduce market discrimination, the wages of black youths have risen relative to those of white youths. Young blacks have made advances in both occupation and education. Yet their *employment* problem has worsened, reaching levels that can only be described as catastrophic.[15]

Freeman and Holzer's book is a report founded on a massive survey, by the National Bureau of Economic Research, of inner-city unemployment, its causes, and possible cures in Boston, Chicago, and Philadelphia. Every effort was made to be sure the sampling was representative of the male population in the most poverty-stricken areas of these cities. And the results? "In 1983 a bare 45 percent of black men who were aged 16 to 21 and out of school were employed, whereas 73 percent of their white counterparts were employed."[16] These researchers wanted to go beyond just the question whether these young black males were working or not, so they made their survey more complicated:

> The topics covered by the survey ranged from standard work activities to the hourly activities of these youths in a day, their desire to work, their use of drugs, their participation in illegal activities, and their perceptions of the labor market. . . .

> A second set of questions . . . dealt [in greater detail] with some of the illegal activities in which the youths might have been involved . . . [and] approximately one-quarter of the total income reported in the sample came from illegal activities. . . .
>
> A third set of questions focused on the activities of these youths in a typical day . . . both productive ones—such as work, searching for work, school and studying, work around the house, and job training—and unproductive ones, such as crime, drug use, and recreational activities (going to movies, listening to music, "hanging out," and so forth). . . .
>
> Another set of questions focused on the attitudes and family backgrounds of the youths. These included the welfare status of their families, churchgoing, residence in public housing, and jobs held by their family members.[17]

Clearly this was more than your typical economic survey of unemployment. Though often couched in the jargon and tabular formulations of the economist, this book gives an interestingly diversified picture of one of the most disadvantaged segments of modern society. Its conclusions are, accordingly, worthy of serious attention:

> A variety of social and economic factors have contributed to the crisis. On the demand side of the market, we find evidence of several determinants, including local labor market conditions and demographics, discriminatory employer behavior, and the unattractive characteristics of the job held. On the supply side of the market, we find aspirations and churchgoing, opportunities for crime, the family's employment and welfare status, education, and the willingness to accept low-wage jobs all to be important factors.[18]

Putting the matter somewhat more colloquially, the studies in *The Black Youth Employment Crisis* offer strong evidence that the jobs available to young black males (not only in the target areas of the NBER survey—inner-city Boston, Chicago, and Philadelphia—but elsewhere as well) are often far from home, and provided by small firms with few checks on the racism of employers or managers, are generally unattractive and in many cases offer fewer rewards (economic or other) than do criminal activities. Further, the young males actually looking for work will be much more likely to find and keep a job if they do not come from a welfare family, have strong motivation to work, are high school graduates, and go to church on a regular basis. In short, the studies reported here support all the worst fears people have about the tragedy of unemployment among young black males.

Is there anything that can be done about this sad situation? Some evidence that people are thinking about this can be seen in "Facing Grim Data on Young Males, Blacks Grope for Ways to End Blight."[19] The main focus of the story was on what was billed as the first national conference on the topic, being held at the time in Kansas City; but there were also references in the article to similar conferences in New York City and Atlanta, and to panels at the national meeting of the NAACP. At the Kansas City conference, a new group was formed, the National Coalition of African American Men, to serve as a research and advocacy organization, among other things to help reshape public policy.

If this organization grows and is effective, it may be the first to succeed where all else has failed. Not that there have not been earlier efforts, from big brother organizations to police athletic leagues, from particular local YMCAs and other church groups to neighborhood houses. But the fact of the matter is that, up to the present, the social agencies a great many of these young people are most likely to encounter are welfare agencies and young offenders programs of the courts. And these most often do little to help young black males find adequate employment.

In short, if there is an area of the breakdown of family life in our technological culture that calls for more activist politicking, it is hard to imagine. One can only wish the National Coalition of African American Men well—and hope the group finds effective ways to work with the NAACP, the National Urban League, and similar activist organizations, including thousands of local community organizations across the country.

I have no national data to fall back on here, but the experiences I have had in Delaware and a few other states suggest that social workers are heavily involved in local programs attempting to improve the plight of unemployed black males. But they, and the agencies they work with, tend to have little power to change an employment picture affected negatively by racism, by distant jobs or jobs that require skills these youths do not have, by the failure of the churches to reach many of these young people, and by disproportionate family breakdowns and female-headed households.

A RESPONSE TO PROFESSIONAL FAMILY INTERVENTION

A strong objection to social work and other interventions in family life has been raised by Brigitte and Peter Berger, in *The War over the Family*. The Bergers say that social work often involves a "bureaucratic reductionism," a "systemic approach" which "has facilitated the notion that the family as an institution is in deep trouble."[20] This they contrast with their own view—they claim it is intermediate between the professionals' view and that of traditionalist defenders of the family—which sees the family as the most important institution mediating between individuals and larger social institutions, including social work agencies.

The Bergers trace the "tension" or "difficult relation" between families and social work professionals "to the beginning of social work, in the nineteenth century." But they say this tension "increased with the ascendancy of professionals since the New Deal legislation and their augmentation through the 'Great Society' programs of the 1960s."[21]

The Bergers admit that, "It is fair to say that all this legislative history has been 'pro-family' in intention." They add immediately, however, that, "intentionally or not, this body of laws (making up, in the aggregate, what we now know as the American welfare state) provided powerful handles for the intervention by professionals in the lives of individual families."[22]

Put more bluntly, this is the too-common complaint that social workers and other family interventionists are "faceless bureaucrats" unresponsive to family members as individual human beings.

THE NEED FOR PROFESSIONAL ACTIVISM

I want to respond to this sort of criticism with a personal note. My wife is a social worker and has been president of the leading social work society in Delaware. She has also been active for a number of years in that society's children, youth, and family issues committee. And she has worked closely with social workers in Philadelphia, throughout southeastern Pennsylvania, and for that matter throughout the country. These people may occasionally worry about "professional imperialism" and unwelcome intrusions by "faceless bureaucrats" into family life, but they are far more likely to be concerned about issues of an entirely different sort.

Almost all social workers seem to be overworked and their agencies understaffed. And low salaries for social work often leads to employing people as social workers who do not have the requisite professional training.

Another concern is over the number of social workers who are abandoning traditional social work in favor of setting up counselling practices. In a recent exchange in *Social Work*, Howard Jacob Karger complained:

> When we sever the historical precedents that distinguish social work from other professions—working with the poor and disenfranchised—we risk becoming the same as psychology and psychiatry practitioners, except we charge less. If we cut our ties to the poor and enter the mainstream of American society, we begin a losing battle.[23]

But the biggest concern by far is the ineffectiveness of professional social work alone, even in the best programs in the best states. An inordinate amount of a social worker's time—especially when compared with the work of any other profession—must be spent in coalitioning, in activist involvement to drum up support for programs, in lobbying with state legislatures and other officials, and in bringing heartbreaking cases to the attention of the public.

This last concern is no complaint—though individual social workers may complain about the amount of activism they must get involved in. Problems of child abuse, of teenage pregnancies, of unemployment among young black males—in short, all the problems of families in our technological society—demand activism. There is no way that professional social work alone—or, for that matter, professional psychology, counselling, youth rehabilitation work, and so on and on—will solve these massive social problems. Indeed, talk of "solving" such problems is probably misguided. All we can realistically hope for is alleviation of some of the worst of the problems—and then only with widespread community support and activism aimed at eliminating some of the broader social, economic, and political conditions that lie at the root of the problems.

The situation, however, is not totally hopeless. There are community groups—large numbers of them, in fact—which take it upon themselves to act as an advocacy constituency in league with the professional social workers and child welfare specialists. The Children's Defense Fund is one of the oldest and strongest. And in 1990 San Francisco hosted the sixth National Conference on Advocacy for Children in the States, which was attended by representatives of hundreds of advocacy groups across the country.

CONCLUDING COMMENT

I come back, then, to the moral I want to draw from this chapter. Social workers traditionally have been among the most activist of the professions—a tradition that could be threatened today. What I argue, in the next several chapters, is that all the technical professions need to follow the example of the social work tradition. Family problems may be no worse in a technological world than they have ever been. Clearly they are no better, hence the need for renewed activist efforts on the part of social workers. However, other social problems of the modern world have clearly been exacerbated by technological development. That means they have been exacerbated by, as indirect effects of, efforts of technical professionals to improve society. (I am not here worried about *misbehaving* professionals.) So, if technical and other professionals have helped to create social problems, it would seem obvious that they also have a responsibility to help solve (or at least alleviate) them. And, if ameliorating social problems requires that professionals go beyond their professional work to join in activist movements—as is obviously the case with social workers and others in the "helping professions"—they would seem to have a responsibility to do that as well.

What I do in parts two and three is see how this plays out in areas of technical professional activity.

Part II

Urgent Social Issues of Our Technological World

4

Technology Educators and Reform

The secondary carriers [of modern consciousness] are processes and institutions that are not themselves concerned with [technological] production but that serve as transmitting agencies for the consciousness derived from this source. The institutions of mass education and mass communication generally may be seen as the most important of these secondary carriers.

—*Peter Berger*

The current reform movement in science education has several fronts. One group [of] . . . proponents of reform see little wrong with the present science curriculum that cannot be dealt with by making courses more rigorous and academic, raising standards, and focusing on "quality" and "excellence." . . .

A second group of reformers see the need to recontextualize and modernize the traditional science curriculum. . . . Scientific Literacy [according to them] needs to be viewed as the major part of civic and social intelligence essential for effective participation in a culture distinguished by achievements in science and technology.

—*Paul D. Hurd*

As I move to consider major social problems of our technological culture, I deliberately begin with a quote from *The Homeless Mind*, by Peter and Brigitte Berger and Hansfried Kellner[1]—to the effect that education plays a key role in socializing young people, poorly or well, into the mindset required for life in a technological world.

Cries of alarm over the crisis in education are as old as the United States, but within the past two decades—as technology has become a major concern in public debate—such cries have become louder and shriller than ever. In 1983, a National Commission on Excellence in Education issued a report with the telling title, *A Nation at Risk: The Imperative for Educational Reform.*[2] Pierre S. DuPont IV, who once said he wanted to be Delaware's "education governor" (and who may have inspired George Bush to promise that he would be an "education President") has now proclaimed the previous decade's reform efforts a failure:

> So for all our time, all our efforts, all of this "cash register mentality" [education expenditures up seventy percent while inflation went up only thirty percent], just what have we accomplished? Sadly, not very much.[3]

In an echo of complaints heard more and more often, DuPont details the evidence:

> Twenty-nine percent of American students who enter the ninth grade fail to graduate from high school. . . . Half of our high school seniors cannot identity Robert E. Lee. . . . The number of high-school seniors scoring above 750 on the combined verbal and math sections of the Scholastic Aptitude Test is 50 percent lower than it was 10 years ago. On international math and science tests, American students score 12th best in the world.[4]

This diatribe appears in the midst of an appeal for what seems to me a misguided "choice" plan that would, among other things, subsidize parents who choose to send their children to private schools in order to make the next generation "truly competitive." Still it does summarize in a pithy way the laments of critics all across the political spectrum.

The outstanding voice in this chorus of lament has been E. D. Hirsch, Jr., especially in *Cultural Literacy*.[5] What Hirsch calls for is the learning of a limited set of those concepts that are the key to "cultural literacy." "This country," he maintains "will not begin to make adequate improvements in our system of education until we reach agreement on a central core of knowledge that should be possessed by all children."[6] This view is popular, and over 2,500 educators are reported to have joined Hirsch's Cultural Literacy Foundation.

Hirsch's book, when it first appeared, was widely viewed as part of the neo-conservative movement in the 1980s. Hirsch admits that he is opposed to inroads of progressive education into the schools over the previous two decades; indeed, he says that is one explanation for the precipitous decline in educational achievement, most notably in science scores, on the part of elementary and secondary schoolchildren. But he says one indication that he is not reactionary is his concern for the education of minorities and other underprivileged children. According to Hirsch, they suffer, to a much greater extent than most, from a knowledge deficit in terms of key concepts needed for a decent life in our technological culture.

In this chapter, I focus on that aspect of the education crisis epitomized in laments over American highschoolers ranking twelfth (that is, near last among the countries surveyed) in their knowledge of math and science.[7] What I am most interested in is the reformers who are attempting to do something about this situation.

THE TECHNOLOGICAL LITERACY THROUGH SCIENCE, TECHNOLOGY AND SOCIETY MOVEMENT

As suggested at the beginning of this chapter (quote from Paul Hurd[8]), there are many sorts of reform proposals out there. One which seems to follow lines Hurd would approve of has been spearheaded by a scientist, Rustum Roy, operating out of the Science, Technology, and Society program at Pennsylvania State University.

A handy summary of this group's efforts in behalf of enhanced science education in the schools is available in *S-STS Reporter* (1985-1988) and its successor, *STS Reporter* (beginning September 1988; the first number of *STS Reporter* includes a list of all the contents of *S-STS Reporter* from 1985 to 1988). These publications document the establishment of a national network, which includes primarily science and social studies teachers for grades seven to twelve. Its goal is to provide STS instructional materials, including bibliographies for all instructional levels, arranged by STS topic. Also provided are short lessons and activities, teacher resources, including lists of STS publications and professional organizations, various states' department of education guidelines, and similar bits of helpful information. The network also provides student modules and reviews books and media aids.

Most important, the national STS network links leaders in the states not only with the headquarters at Penn State but with one another.[9]

Illustrative of this groups's approach are the criteria used in evaluating teaching materials for class work:

1. Are the relations of technological or scientific developments to socially-related issues made clearly, early, and in ways that capture attention?

2. Are the mutual influences of "technology," "science," and "society" on each other clearly presented?

3. Does the material develop learners' understanding of themselves as interdependent members of society and responsible agents within the ecosystem of nature?

4. Does the material present a balance of viewpoints about the issues?

5. Does the material help learners to venture beyond the specific subject matter to broader considerations of science, technology, and society, including the treatment of personal and social values/ethics?

6. Does the material engage students in developing problem-solving and decision-making skills?

7. Does the material encourage learners to become involved in a group or personal course of action?

8. Does the material use the STS linkage to foster learners' confidence in handling and understanding at least one limited "science-

technology" area, and/or handling and using some quantification as an aid to judgments in the STS area?[10]

The STS network has been closely associated with a series of technological literacy conferences (the seventh was held in 1992), and Rustum Roy was the moving force in the formation of the National Association for Science, Technology, and Society in 1989.

Each year it becomes more obvious that STS as an approach to technological literacy in the schools is catching on. However, it is still much too early to tell whether or not the movement can be sustained—and, even more importantly, whether it will have a genuine impact on the scientific or technical literacy of the next generation of high school graduates.

THE CRISIS OF HIGHER EDUCATION

If there are laments over the way science has been taught at the elementary and secondary school level—and over an almost total lack of teaching aimed at *technological* literacy—this negative chorus has recently turned against colleges and universities as well. Paul Hurd, in fact, blames most of the problems at the lower levels on bad teaching at the college level: "Another obstacle to the development of scientific literacy is the way teachers are educated in science departments"—namely, in courses in which "science lessons consist primarily of vocabulary drills reinforced by tests of the same nature."[11]

Criticisms of higher education, however, go far beyond just concerns over science teaching that do not encourage undergraduates to choose science or science teaching as an exciting career. Here again the chorus of criticisms is as old as the republic, but it has reached a crescendo in the past decade—alongside the chorus of criticisms of the schools. Derek Bok, past president of Harvard University, has recently published an interesting version of the new critique. One point Bok stresses is the failure of our colleges and universities to do enough to help society deal with pressing social issues of the day.[12] Ernest L. Boyer, president of the Carnegie Foundation for the Advancement of Learning (a long-time source of criticisms of American education), has taken the opportunity provided by the publication of Bok's book to issue his own Bok-like prescription:

> The application of knowledge to contemporary social and civic concerns can and should be a more highly valued part of scholarly endeavor. What's needed on American campuses today is a recognition that the application of knowledge [as well as its generation] is scholarly work that flows out of serious inquiry.[13]

Although I share some of these broader concerns, I limit myself here to the technological literacy reformers rather than the broader issues of relevancy raised by Bok and Boyer. It seems to me—as I will say at the end—that an effective answer to the broader questions may depend, at least in part, on the right kind of resolution of the science, technology, and society issue.

One prominent group of reformers who have emphasized college training in technological literacy is the Council for the Understanding of Technology in Human Affairs. This group—which has now gone out of business, somewhat surprisingly taking its mission to have been accomplished—proclaimed that it would foster "collaboration among engineering and liberal arts educators," that it would enlist "the participation of representatives of industry, government, professional organizations, and public interest groups concerned about the role of technology in human affairs."[14] CUTHA, as the group called itself, worked closely for over a decade with the Alfred P. Sloan Foundation's New Liberal Arts Program in an effort to introduce technological literacy into liberal arts colleges.

The effort led to a New Liberal Arts book series. The foreword to early books in the series[15] includes this statement:

> The Alfred P. Sloan Foundation's New Liberal Arts (NLA) Program stems from the belief that a liberal education for our time should involve undergraduates in meaningful experiences with technology and with quantitative approaches to problem solving in a wide range of subjects and fields. Students should understand not only fundamental concepts of technology and how structures and machines function, but also the scientific and cultural settings within which engineers work, and the impacts (positive and negative) of technology and society.[16]

CUTHA and NLA are seen by many as fairly defensive, as hostile to critics of technology. But they do represent an instance of strong advocacy on the part of a technical community that has, historically, been reluctant to get involved in activism.[17]

These efforts, predictably, have been criticized; critics of technology tend, for example, to see the spokespersons for CUTHA and NLA as advocates of technological development at the expense of the environment; as advocates of technocracy; even as advocates of technological imperialism. Moreover, some engineering educators, who might be expected to be well disposed, are also critical of CUTHA and NLA. A major criticism is that NLA-based courses tend to introduce students to science or quantitative thinking in general rather than to engineering methods or technical decisions as they must be made in the real world. Another criticism, aimed not at NLA but at defenders of the STS approach, is that it is too vague and poorly defined. Still another criticism is that neither NLA nor STS has managed to have any real impact on undergraduate education. At best, where NLA or STS programs or courses have been started, they remain peripheral, dependent exclusively on outside funding. They are, it is said, havens for "soft" elder statesmen from science or engineering departments rather than places where younger research professors make a name for themselves.[18]

RESPONDING TO THE CRISIS OF HIGHER EDUCATION

The critics have a point—as do the technocratic reformers of higher education. My concern here is with how *real* reform can be brought about. And it seems to me that two things are missing on both sides of the issue as discussed so far. These are, first, education for citizenship, and, second, the activism needed to bring about genuine reform.

The one group of academic reformers who seem to have anything serious to say about preparation for citizenship are those who focus on *interdisciplinarity*—especially in the humanities.[19] STS and NLA spokespersons often talk about education for citizenship, but that seems most often to translate into a concern that people at the beginning of the twenty-first century should acquire technical knowledge if they are to exercise the role of enlightened citizenship in a technological world. It is not typically a concern over what citizenship itself means and how one learns to practice it.

A hundred years ago, when professional education was beginning in the United States, the new professional fields had to earn entry into the universities by proving that their future practitioners could benefit from a solid grounding in the liberal arts. Today, concerns of this sort have very nearly disappeared from higher education.[20] They have been replaced, if at all, by such things as Harvard's new core curriculum, which has been emulated in dozens of ways by hundreds of colleges in a period of slightly more than a decade. Some of the designers of these core programs recognize that the only hope we have today of grounding career training in the liberal arts is in humanities-based interdisciplinary programs.[21]

Why is this? For years, Julie Thompson Klein has been collecting samples of the rhetoric used by interdisciplinarians in defense of their ideal. Interdisciplinarity, Klein reports, is said to be an instrument of "liberation," "convergence," "integration," "symbiosis." Similarly, in their attacks on discipline-based university departments, Klein shows, the interdisciplinarians claim that disciplinarians are simply defending turf or empires—or, at worst, that they are "authoritatively performing mortuary rites over corpses of dead bodies of knowledge." In summary, Klein says that, whether interdisciplinarians unite under the slogan of "system" or that of "organism," in either case, "It is very clear that the dominant conception of interdisciplinarity is that of a productive art of restoring and discovering the grounds for interdependence and relationship."[22]

This is the key. Without some ability to integrate, to relate parts of their education to one another and to the world outside the academy, graduates of even the best universities would be no more than well trained technicians. This is not to disparage technicians; they become more important every day in our technological world. But technical work requires leadership, coordination, direction—even vision, I would say. Where might we get that today if not from university graduates?

Klein went on later, in *Interdisciplinarity*, to provide a truly comprehensive survey of interdisciplinary research and education. There, in a capstone chapter and conclusion, she surveys almost all the efforts to provide an interdisciplinary core for undergraduate education in the United States—and even several crucial efforts from other countries, including Great Britain, Australia, Denmark, Norway, and Japan. Her conclusion, based in part on a 1986 survey by William H. Newell for the Association for Integrative Studies, is that there has been an "interdisciplinary renaissance" in the United States. It is, she says, "linked strongly with the desire to revitalize the core of the liberal arts." Newell's survey turned up 235 programs, most of them relatively recent and concentrated in state universities and community colleges—though there are also interesting examples at Brown, M.I.T., and Stanford, among other recent programs, and also the long-established programs at Columbia, the University of Chicago, and St. John's College in Annapolis.[23]

On the other hand, Klein also concludes that there is as much reason for pessimism as for optimism: "Interdisciplinary programs have been limited in three major ways: by the lack of a long-standing tradition for interdisciplinary education, by the power of disciplinary and departmental boundaries, and by the influence of conditions outside the universities"—such as economic recessions and diminished public support for higher education. Under this pessimistic head, she concludes: "The experiments of the 1960s and early 1970s have not supplanted the disciplines, and their 'instrumental' orientation has directed attention from the broader 'synoptic' concerns of interdisciplinary work."[24]

The urgency of the need for interdisciplinary problem solving in a technological world has been pointed out many times, but I find this call to action particularly compelling:

> Reasonable criticisms of technological development, of particular technological innovations, of their goals and consequences, issues of values and moral responsibility as these impinge on the socio-cultural situation of technology—all these are the prerogative not only of humanists but also of social scientists and politicians, of reflective planners and systems analysts, of the technological experts themselves. Such problems urgently demand solutions adequate to their systemic character.[25]

This plea, from German philosopher Hans Lenk, is aimed principally at other philosophers, and surely it is an important point for philosophers interested in technology to heed. Moreover, for those who take an activist approach, the need for a reform of higher education—the need to give it focus and direction—seems to be a major social problem of great urgency.

Perhaps so, others would say, but of what use is academic scholarship? If we make the traditional distinction between scholarship and practically-oriented scientific/technological research, many critics would say it is of no use; it is idle curiosity, a means of self-aggrandizement for learned dons with esoteric interests. Defenders of the traditional learned disciplines, on the other hand, retort that the arts and humanities are the last, best hope of civilization in a world given over almost entirely to the pursuit of power, including technological power.

I do not wish, here, to take either side in this dispute, but it should be clear by now that I am concerned with the solution of social problems. In my view, much that passes for scholarship in the universities, in technical fields as well as in the traditional arts and humanities, is merely technical and does not meet the needs of society—either the need for practical solutions or the need for vision. This seems to me the unspoken complaint that lies behind much of the rhetoric of interdisciplinarity reported by Klein. It may be that the universities could go on for a long time doing what they have been doing for the past half century. But it seems unlikely that even the leaders of the political and corporate power structure—let alone the activist reformers to whom I make my appeal—would tolerate that state of affairs forever. In my view, both a fragmented university of non-communicating specialized professional communities and a university turning out specialized technicians for careers in high-technology industry and government agencies represent a serious threat to the traditional ideal of the university.

Whether or not the old ideal is still feasible, the traditional university was once expected to serve the human community in more ways than by generating esoteric journal articles and training career-oriented technicians. I would say that its primary mission was *cultural*. In these terms, the battle for interdisciplinarity, for the integration of knowledge, for wisdom, is extremely urgent. *If it is still possible*, we need a crusade to restore the traditional but ever new ideal of the university as a community of scholars serving especially the cultural needs of humankind.

Is it possible to achieve these lofty goals? And, especially, is it possible without a wholesale transformation of education from kindergarten on? The evidence so far, from efforts to teach science in early grades through STS programs, or from efforts to create interdisciplinary cores in colleges and universities, would suggest that at best it is going to be an uphill battle. I would say, moreover, that the situation is not likely to change very much unless there is a redirection not much talked about heretofore. So far, educational reformers have operated mostly within the educational system. There has not been citizen activism at a level necessary to bring about the desired changes, even where publicity ventures have been mounted to awaken the general public to the enormity of the problems. In my view, educational reformers will never succeed without an enlightened public constituency for educational change—especially change of a fundamental sort. It may be too much to ask educational reformers to go out and create a constituency for change. But if they do not, and if the right sort of constituency does not emerge on its own or as a result of other historical pressures, it seems likely that basic educational reform—whether in technological literacy or in education for leadership—will remain largely peripheral. Educational establishments, at all levels, may adopt the new programs for a while; they may even, in some cases, mount major crusades for reform. But demands for technological literacy on the part of all (or most) citizens, for humane and socially responsible engineers and other technical professionals, and for an integrative core of humanities-based studies aimed at leadership are likely to continue to seem tangential or peripheral to the main task of pushing a majority of children through secondary school and of preparing the college-age population for technical careers.

That much said by way of pessimism, I still hope that the reformers can succeed—that a constituency for fundamental educational reform can emerge and will work to support the educational reformers in their best efforts.

5

Medical Educators, Technology, and Health Reform

The education of new doctors has not prepared them for management roles. . . . The rapid advancement of medical technology has dramatically altered patient care patterns, as have the determined efforts of third-party payers to seek more value for their increasing health care investments.

—William B. Walsh

A special feature of medical specialization and technological innovation is that the two are simultaneously parallel and interactive, creating an impetus to further technological innovation and specialization. . . . [This spiral] results in increasingly sophisticated medical specialties.

—Anselm Strauss

The medical establishment in the United States has been under attack from radical critics for well over a generation. In 1970, a group calling themselves the Health-PAC wrote a scathing attack on the bureaucracy and inaccessibility to patients, especially the poor, of our hospital-based health care system, under the title, *The American Health Empire*.[1] In 1976, Ivan Illich, in *Medical Nemesis*,[2] blasted the health establishment for systematically depriving people of their traditional ability to control their own health care. In the decades between then and now, a groundswell of popular criticism has arisen, culminating in the grassroots "health decisions" movement of the 1980s.[3] By the 1990s, even large corporations had begun a serious re-examination of health costs, trying wherever possible (for example, in labor negotiations) to shift the burden of health care insurance to workers. (This movement came, somewhat ironically, at the same time that spokespersons for the business sector were touting pharmaceutical and other health-related stocks as the best investments for the future.) All of this ferment can be summed up most simply by saying that the crisis in health care in the United States is widely perceived to be among the worst problems of our technological society.

Reformers in medical education are sympathetic with these cries of alarm over health care, and a number of significant calls for the reform of medical education have been made in recent years. What I find missing in them is provision for the right kind of education—namely, education in *social responsibility*. My topic in this chapter is the impact of high-technology medicine on medical education. More particularly, what I focus on is recent proposals for the reform of medical education, the prospects for health reform in an era of technology-driven specialization, and the relationship between the medical humanities and the right sort of reform. What I claim is that medical education (and the corresponding training of nurses, physical therapists, medical social workers, etc.) needs to prepare health professionals as much for the exercise of social responsibility as for the exercise of expert decision making.

Most of today's medical education reformers focus on the novelty of health care arrangements today, on issues of third-party constraints on practice, and so on; occasionally, they also lament the imbalance between students choosing greater and greater specialization and the urgent needs of society for more primary care providers. All the while, the pace of new inventions of high-technology medical equipment quickens, along with a constant flow of new drugs and medications, and medical students are as quick as any other students in seeing the career advantages of knowing how to manage these new scientific and technological discoveries—which they and their professors can, at the same time, readily rationalize as leading to, in the popular slogan, "the finest health care in the world."

The only medical education reformers I am aware of who go against this flow—and who have at least some chance of preparing medical students and other health professionals to shoulder social responsibilities as part of their professional commitment—are the small number who are pushing to include education in the medical humanities as an integral component of health care education.

BACKGROUND: THE CHANGING HEALTH CARE SCENE

Some of the medical education reformers I will refer to base the urgency of their calls for reform on "our changing times." An example is William B. Walsh, already cited in the headnote above. He is introducing one of the better known proposals, "The Reform of Medical Education." Walsh complains that:

> The education of new doctors has not prepared them for management roles, with the result that destructive conflicts arise with administrators who are better equipped to oversee the nonclinical business dimensions of these arrangements, but who have no capacity to treat patients. The rapid advancement of medical technology has dramatically altered patient care

> patterns, as have the determined efforts of third-party payers to seek more value for their increasing health care investments.[4]

Of the three aspects of our "changing times" mentioned here—new management roles, advances in medical technology, and institutional pressures—the one I want to focus on is medical technology, including new drugs, new prosthetic advances, and the like.

I am impressed with the account of these changes in *Social Organization of Medical Work*, by Anselm Strauss and colleagues. They preface their work with this succinct judgment: "The work of physicians, nurses, and associated technicians has been radically and irrevocably altered by today's prevalence of chronic illnesses—the illnesses that bring patients into contemporary hospitals—and by the technologies developed to manage them."[5]

Strauss and his co-workers then stress the overwhelming impact of medical technology on contemporary hospitals:

> The diagnosis and treatment of the chronic illnesses have contributed to the widespread use of a great array of drugs, rapidly increasing numbers and types of machinery . . . and, of course, various surgical and other procedures.[6]

This is related, they say, to "a considerable industry [that] has evolved for manufacturing and supplying drugs, machines, and other elements of these technologies." As a result:

> New occupations are growing up around the servicing and utilization of this machinery (bioengineers, safety engineers, respiratory therapists, physiotherapists, radiology technicians), and many of the medical specialties are centrally dependent on its [the machinery's] use.[7]

It is in this context that Strauss and colleagues make their claim about a spiral of technological innovations and technology-related specializations.[8]

It seems to me that the accompanying bureaucratization of health facilities is one cause of the dehumanization that critics lament so forcefully. Another effect is a new set of roles for health workers that are needed to remedy the negative effects of fragmented and machine-driven care.

The Social Organization of Medical Work includes a horrendous example involving nurses. The authors cite field notes:

> I was following the resident who entered a woman's room, accompanied by five or six medical students, to do a cervical examination. They all stood closely observing the resident do his examination of her vagina. It was done with scarcely an introduction, with little explanation or attention paid to her reactions. A nurse did hold the patient's hand. The resident did vaginal scraping, demonstrating to the students how to do this right. Then they all

exited without a single word, only a passing nod from the resident. After they left, the patient burst into sobs, while the nurse consoled her.

They then comment:

> Such rectification work can include apologetic or caustic remarks about the brusqueness, inconsiderateness, or callousness of the offending staff member. [But] nurses [also] often do this rectification after patients have been visited by technicians or after they have visited machine sites for diagnosis or therapy.[9]

The nurse's "new" role here may seem rather traditional, but we should not forget that nurses, too, are today faced with ever more sophisticated technological roles.[10]

PROPOSALS FOR THE REFORM OF MEDICAL EDUCATION

Against this background, I here sketch out briefly the suggestions of three of the best-known recent proposals for reform in medical education.

The most comprehensive reform proposal is "Physicians for the Twenty-First Century,"[11] the report of the Project on the General Professional Education of the Physician (GPEP) of the Association of American Medical Colleges. It is organized around a set of five conclusions, dealing with the general purpose of professional education, undergraduate preparation (viewed as an essential part of) for medical education, learning skills, clinical education, and faculty involvement. Several recommendations are made under each heading—for instance, that medical education should emphasize skills, values, and attitudes as much as the acquisition of knowledge; that future physicians should be prepared to deal with changes occurring in the health care system; and that physicians should be taught to work not only with individual patients but with communities in health-promotion and disease-prevention programs (these three under the first or general heading).

With respect to pre-med education, the report endorses a core of broad liberal arts requirements, communication and writing skills, and only "essential" pre-med courses.

Within the undergraduate medical curriculum, the proposal endorses independent learning, fewer lecture hours, and the use of new problem solving and information science techniques. For the clinical years, GPEP not only recommends integrating clinical education with the basic sciences but also discourages students from spending excessive time preparing for residency positions.

Finally, under faculty involvement, the report recommends coordination, leadership, and the mentoring role. As an addendum, under an "other" heading, GPEP acknowledges, among other things, the need for equity of access for minority students.

A second influential medical education reform proposal has been put forth by Robert H. Ebert, former dean of Harvard Medical School, and Eli Ginzberg, an economist from Columbia University.[12] This proposal was supported by the Robert Wood Johnson Foundation. Ebert and Ginzberg, like the GPEP authors, organize their reform proposal so as to justify a set of recommendations: for example, to combine the last two years of medical school with the first two years of graduate medical education; to adopt more flexible admissions policies, including early admissions and larger numbers of minority students (federally funded if need be); to encourage M.D.-Ph.D. research programs; and to look for better ways of supporting graduate medical education.

This special issue of *Health Affairs* also includes comments from various constituencies, including an interview conducted by the editor, John K. Iglehart, with Robert G. Petersdorf of the Association of American Medical Colleges;[13] the editors headline this interview as "Medical Schools and the Public Interest," and, in effect, it constitutes a complementary proposal to that of Ebert and Ginsberg—with the main emphasis on improvements in clinical teaching (but not necessarily in the Ebert-Ginzberg fashion of combining undergraduate and graduate medical education) and monitoring the quality of care (though not in the fashion of the Reagan Administration).

Still another reform proposal, "Clinical Education and the Doctor of Tomorrow," emanated from a Josiah Macy, Jr., Foundation National Seminar on Medical Education and has been summarized by David E. Rogers.[14] Rogers's recommendations include: (1) centralize control of the curriculum, (2) make residency programs the responsibility of medical schools, (3) facilitate educational innovation (including changes in national board exams and the grading of medical college admission tests), (4) move more clinical training to ambulatory settings, (5) require community service as part of training, and (6) require medical students to pass performance-based clinical exams.

These and other reform proposals are related to criticisms that contemporary hospital-centered health care has gotten too expensive and too dehumanized, and to reform movements within the medical education community itself. These have, in many cases, been centered in departments or programs or special projects established in medical schools to evaluate and plan for the future of medical education—supposedly at least in part in response to complaints from medical students and residents.[15]

A MEDICAL HUMANITIES REFORM MOVEMENT

I think this brief survey shows that the one thing noticeably missing in all the medical education reform proposals is any suggestion that the humanities ought to be an integral part of medical education—anywhere, that is, except in a broadened pre-med program. Yet there is a widespread belief, in certain circles, that many of the ills currently afflicting medical education could be cured if the pace were slowed down, if medical school students and residents could be given more time and motivation to *reflect* (in the fashion normally associated with the traditional humanities). As it is now, medical school students and residents must frantically absorb enormous bodies of knowledge and hectically learn to deal with clinical emergencies in a pressure-cooker setting. If the pace of medical school could be made more leisurely, there might be some hope that a new generation of differently-trained physicians could bring different and wiser attitudes to their future practice. A corollary of this might be that we could draw an entirely different type of student into medicine in the first place.

I begin here by citing some of the well known claims made for the role of the humanities in medical education:

J. David Newell: "There are no doubt those who would be quick to say that if [something] has to do with the humanities it has nothing to do with medicine and vice versa. We feel that this [is] a prejudice. . . . Its roots lie in a limited understanding of medicine as much as in an ignorance of the humanities. . . . [There is] an ancient and fundamental conviction that medicine is much, much more than a science; it is at the same time an art. . . . [And] it is an art firmly grounded [both] in the sciences [and in] a fundamental focus on the *human*."[16]

Eric Cassell: "The humanities have always had a place in medicine, and . . . they will play an increasingly important, necessary, and specific role as medicine evolves beyond its present romance with technology toward a more balanced view of the origin and treatment of illness."[17]

Joanne Trautmann: "The study of literature enables medical students to feel that they are more than mechanics of the body."[18]

Alton I. Sutnick: "The study of the humanities enhances the medical student's appreciation of the value of life, without which the physician would have no reason to carry out his or her function. It helps to produce a physician who recognizes the relationship between medicine and other aspects of society, and who is a complete and multifaceted human being able to interact with peers, patients, and community."[19]

John J. Sorenson: "Sometimes we reduce medical education to reflexive thinking; we think students should immediately be able to spout out what the 'correct' answer is. The problem is that the correct answer may change. The humanities, on the other hand, put a high priority on reflective thinking—pondering the possibilities, weighing the alternatives, considering other points of view."[20]

Someone might object that all of this is simply pious talk, with absolutely no real-life impact on medical education as it is typically experienced in U.S. medical schools. That may indeed be true;

however, this says nothing about the force of the claims being put forward by these advocates of the medical humanities.

In fact, the medical humanities have achieved a limited but tenuous toe-hold in medical schools. The oldest programs seem to have come into existence in the 1960s. There are many kinds of programs, and contrasting approaches reveal interesting differences. Some are exclusively ethics programs, typically offering the medical student an opportunity for a different perspective but no degree. Others are broader, with courses in medical humanities in addition to medical ethics—but still with no degree option. Finally, a few programs are both broad, offering ethics plus other humanities courses, and give the medical student a chance to earn a second master's degree or even a Ph.D. in addition to an M.D. Anecdotal evidence suggests that the number of ethics courses in medical schools has increased tremendously in the last decade or so, and some people even claim that up to eighty percent of medical schools have at least one ethics course. As for the medical humanities more generally, the record is much more spotty.[21]

There are even medical reformers who begin to see the need for more radical changes in these directions. Stephen Abrahamson has recently called for structural changes in medical schools—for example, the establishment of deans for educational affairs and vice chairs for education in departments, both with real budgetary control over and a commitment to the genuine education of medical students (as opposed to training for work in high-technology specialties).[22] Abrahamson's brief falls short of making the humanities an integral part of undergraduate medical education, but his position is open to a move in that direction.

Abrahamson's plea is based in large part on the research of Samuel Bloom, a medical sociologist who has attempted to explain why decades of medical school reform proposals have led to so little change. Bloom summarizes:

> To analyse this history of reform without change, this paper first establishes what the content and structure of medical education is, and how it came to be that way; second it traces a process whereby the scientific mission of academic medicine has crowded out its social responsibility to train for society's most basic healthcare delivery needs.[23]

Bloom concludes on a very pessimistic note, though it is also a note that may signal a possibility of basic reform:

> The choice is clearly trending away from people-centered practice and toward the role of technical-specialist. If this observation is accurate, the explanation is not to be found in the motivation or the selection of recruits to the profession. It is in the structure of the situation of modern medicine and in the structure of its major institutions. That is where change must occur if we are not content with the way things are.[24]

Bloom is no clearer than Abrahamson that a medical education including the humanities as a central part of the curriculum might provide more reason for optimism.

The one medical education reform manifesto I am aware of that does refer to the humanities as having a central role to play is Kerr White's *The Task of Medicine: Dialogue at Wickenburg.*[25] In that report of a conference arranged to discuss ideas of Charles Odegaard in *Dear Doctor: A Personal Letter to a Physician,*[26] White talks about "the literate physician," and urges the Association of American Medical Colleges to prepare bibliographies on:

> the history of ideas in science, medicine and health services; evolution of the scientific method; biographies of physicians; health and disease in fiction, art, drama and poetry; physicians as artists and authors.[27]

White is also impressed with the evidence that suggests that medical students who majored in the humanities or social sciences do as well in medical school as those who major in the natural sciences. White, however, seems to share with the establishment medical reformers cited earlier the view that a renewed emphasis on the humanities belongs in the pre-medical program rather than in medical school itself.

Charles Odegaard, in a paper in *The Task of Medicine* that seems to have been the keynote address at the conference, comes much closer to seeing the humanities as central to the education of a caring physician:

> In addition to the development of an *attitude* appropriate to the role of a physician and the acquisition of *knowledge* about the diversity of human nature and the experience of man, the physician also needs to develop *behaviors, skills* in using himself or herself in personal relations with patients and their families.[28]

Odegaard concludes with an explicit claim that humanists and social scientists can contribute to the acquisition of these attitudes, skills, and knowledge; however, he notes, "The physician's purpose [in this] is . . . to use all these sources of knowledge of man and nature as they contain information relevant to his role as a caretaker of the sick"—rather than to become a humanist or social scientist.

A PERSONAL REACTION TO THESE REFORM PROPOSALS

I count myself a student of the medical humanities, and I am currently a visiting scholar in a medical school. However, I want to offer my personal reflections primarily as a philosopher of technology interested in the social responsibilities of scientists, engineers, physicians, and other technical personnel.

We have just seen it argued, e.g., by Bloom, that reforms in medical education are not enough; broad social changes are needed as well.

Some philosophers of technology (e.g., Jacques Ellul)—if they were to comment on medical education—would say there is absolutely no hope that reform efforts of any kind, professional or humanistic, will have even the slightest impact on the ever-increasing technicization of medical education. Others (e.g., David Noble) might say it is hopeless unless health care workers take control of their work lives and resist the introduction of new technologies in health care settings. Still others (e.g., Herbert Marcuse) would be profoundly pessimistic about the possibility that health care workers can develop the kind of revolutionary consciousness, through reformed medical education (or nursing education, etc.), that would be necessary for them to take control in this fashion. Indeed, many critics of high-technology medicine maintain that increasing technicization and specialization of health care—with consequent bureaucratization and dehumanization—are inevitable.[29]

In my opinion, these philosophical antitechnologists and radical critics of contemporary medicine are too pessimistic—though they make good points that should be kept in mind in making more optimistic assessments of the situation. Indeed, that is the only reason for bringing them into this discussion at all. I want now, briefly, to mention the change agents that might help to ground such optimism as is called for.

Most of the medical school reformers seem to me very unlikely to bring about any basic changes in the pattern of students' ever-more-specialized career choices. For the most part, medical school professors or administrators, they would seem almost by definition to be as deaf to the laments of antitechnology radicals as their colleagues are to the cries for reform of medical radicals, from Ivan Illich to Vicente Navarro to the Health-PAC activists of the early 1970s.[30] The reformers occasionally pay lip service to the urgent social need for more general practitioners—or, even more rarely, to the need for primary care provided by health practitioners other than physicians. But the specific recommendations in their proposals genuinely have to do either with minor tinkering with the system or with how ever-more-expensive medical education is going to be paid for.

The medical humanities reformers, perhaps because they are not so centrally situated within the medical education mainstream, might seem to have a somewhat greater chance of success in bringing about change (assuming, of course, that the medical humanities were to become central to medical and health care education). Without a doubt, a more reflective, slower-paced approach to medical education would in itself be a big change. And if there were a widespread change among potential pre-med students in the college years in perceptions of what a more humanistic medical school education would demand, the pool of medical school students might drastically change.

However, this is where the warnings of the extreme antitechnologists can help us to be wary about making any promises of easy success. The "medical school grind" mindset seems firmly fixed in the minds not only of students but of professors—often including the reformers. The ideology of progress in medical *science*, aided by ever-new pharmacological discoveries and technological inventions, remains extremely powerful—and that in the minds of the public as well as the health care community. And it is largely this ideology that fuels the rat race of unreformed medical education and the clamor to get into the specialties that prepare one to use the latest drugs and technologies to provide "the best medical care in the world."

This leads me to make a somewhat provocative claim, that the only way the medical humanities reformers can have any hope of success is by way of activist coalitions with two groups: with consumer groups demanding change in the delivery of health care, and with those people in medical education whose consciences have moved them to be receptive, in the last decade or so, to the introduction of medical ethics or medical humanities into the medical school curriculum.

I need to say a word here about the consumer groups which today are demanding change in the health care delivery system. Bruce Jennings has emerged as the chronicler of what he calls the "community health decisions" movement. Introducing a set of reports on well organized groups in Oregon, Vermont, New Jersey, and California—as well as the would-be umbrella group, American Health Decisions—Jennings says the movement is "alive and kicking." Ralph Crawshaw, of American Health Decisions, says he sees the movement as "nurturing a grassroots philosophy." Jennings himself is somewhat skeptical, worrying that some of the rationing these grassroots movements are fostering may "not be very equitable" and may reflect the middle class biases of most of the activists in the groups. But Jennings remains optimistic and hopes that these movements, now active in more than a dozen states, will

> prove the conventional wisdom wrong, and . . . discover that the transformation from a private to a civic outlook is easier, and more urgently desired, among the American people than we thought.[31]

What Jennings has in mind here is a hope that the grassroots health movements will spur a broader constituency than just the middle class to get involved. As a result, a broad constituency for change might emerge that might be more equitable, more concerned about the health needs of all citizens.

CONCLUSION

Could a coalition of grassroots activists, humanities reformers, and medical education reformers succeed in redirecting medical education toward greater social responsibility? There may be some hope.

First, the health consumer activist movement, though it is only beginning to be visible, seems to have the potential to tap a great groundswell of public opinion critical of current expensive and inaccessible high-tech medicine. (However, a warning is in order: as is almost always the case with public opinion, there is a great deal of ambiguity here; an aroused public may well demand, at the same time, both more humane health care and the latest in medical gadgetry and "miracle cures.")

Second, the medical humanities reformers have both a formal organization, the Society for Health and Human Values, and a toe-hold in at least some medical schools. What they seem often to lack is both a sense of identity as reformers and the sort of boldness that might lead them to make stronger claims in the medical schools and to join with the health consumer activists.

Third, the best-situated of all are those medical school faculty and administrators who already recognize the need for education in medical ethics and social responsibility. They have already successfully introduced medical ethics (along with nursing ethics, etc.) into the schools. Some have even introduced medical humanities more generally. What they have not succeeded at is making the medical humanities, and social responsibility concerns, a central focus of medical education. There is also a question about how likely they are to continue their support if the medical humanities reformers do make common cause with the health consumer activists.

In sum, is there any prospect that the drive toward ever more complex technologies along with the continuing tendency of medical students to specialize in related fields can be slowed or even stopped? While extreme pessimists would say no, I say there is a chance. But there is a chance only if the would-be reformers recognize how much is at stake and take the appropriate activist measures.

6

Media Professionals and Politics

Through school curricula, motion pictures and television, advertising of all sorts, and so on, the population is continuously bombarded with ideas, imagery and models of conduct that are intrinsically connected with technological production. . . . [And] these themes become incorporated in a modern world view.

—Peter Berger

What is the current effect of technology on children and adults, and what kind of person is being created by the millions, right now, without the least genetic intervention? I would describe this person, as I have encountered him in others and in myself, as a captivated, deluded, and distracted individual. In our society, having once been obsessed with work, we have now become fascinated by the multiplication of images, the intensity of noise, and the spread of information . . . [especially] in television and mass entertainment. There is no exit.

—Jacques Ellul

I begin again with Peter Berger's characterization of a carrier of "modern consciousness"—this time, the media.[1] In this book I am arguing that, as an antidote for the social ills of technological society, a progressive political movement or movements is better than (among others) radical proposals that call for putting technology in the hands of workers. Political activism, however, clearly presupposes that the political process has not been so corrupted— radicals would say by technology-based corporate capitalism using the media[2]—that activism is pointless.

I agree with those who say that democracy in the United States today is troubled. Just before the 1990 elections, the *New York Times* ran a story under the heading, "Unhappy Voters Notwithstanding, Few Incumbents Seem in Danger."[3] What this story suggests is that the old saw, "If you don't like what politicians are doing, throw the rascals out," has become almost impossible to implement in the United States today. Michael Oreskes, the reporter writing this story, says: "Most lawmakers are certain of re-election, either because they are unopposed in the balloting on Nov. 6 or they face an opponent so underfinanced as to be out of contention." What has brought about this sad state of affairs? Many critics point to problems in redistricting; after each decennial census, incumbents of both major political parties redraw district lines in a way that helps ensure their re-election. Other critics focus on the negative influence of money on elections—money from political action committees (PACs), for instance, goes overwhelmingly to incumbents, and particularly to those who head powerful committees in Congress or in the state legislatures. Still others focus on new technology-related causes. One example is computer-assisted voter targeting, which concentrates most media coverage on swing voters in especially important districts—often less than five percent of the electorate, typically white or better educated than the average voter or otherwise singled out for attention. The other technology-related influence is the media, and most especially television.

In this chapter, I consider three sorts of worries that critics have expressed about the corruption of the democratic process by technology—especially the technologies just mentioned, which are normally lumped under the heading of "the media."

TECHNOGIMMICKRY AND DEMOCRACY

The first sort of criticism has been getting louder and shriller in the United States in conjunction with the presidential campaigns of 1980, 1984, and especially 1988. However, concerns over the possibly corrupting influence of television on the electoral process have been voiced at least since the Kennedy-Nixon debates of 1960. The concern now is not just about TV, but about specialized polling techniques, about targeting of particular audiences by means of cable TV or computerized precinct mapping, about "focus groups" and other psychological profiling of voters, and similar tricks of the professional campaign manager's trade.

A useful collection of expressions of concerns of this sort can be found in a volume edited by Joel Swerdlow, *Media Technology and the Vote*.[4] There such critics of the media as Newton Minow, Julie Kosterlitz, Jeff Greenfield and political commentators such as George Will and Jonathan Schell (with a long list of others) address pros and cons of a broad range of issues. Should government regulate TV? What are the effects of TV on the costs of campaigns? What are the implications of computer-driven segmentation of the electorate? What are the effects of TV on the political parties? or on voter turnout? And so on.

After the largely negative presidential race of 1988, several books have appeared. One title tells a great deal: *See How They Run: Electing the President in an Age of Mediaocracy* (1990);[5] it is by Paul Taylor, a political reporter for the Washington Post. The book addresses such questions as "why should politicians talk about issues when thirty-second 'sound bite' advertising is so effective? or Why not run negative campaigns when there is so much that is negative for diligent investigative reporters to uncover?"

Nicholas Lemann, himself a well known journalist who often covers electoral politics, has reviewed *Road Show: In America, Anyone Can Become President; It's One of the Risks We Take* (1990),[6] by Roger Simon, a political columnist and reporter for the *Baltimore Sun*. The *New York Times Book Review* editors gave Lemann's review the telling title: "How the Spin Doctors Operate."[7]

But Lemann also (perhaps inadvertently) suggests the limitations of the claim that politics is being corrupted by technogimmickry:

> The most optimistic thought about the campaign system that comes to mind from Mr. Simon's description of it is that we might be in a temporary crisis that exists because most of the people now of Presidential-candidate age are too old to have learned their political skills in the television era. . . . We can hope . . . that aspiring politicians of the future will learn to master television fairly effortlessly as part of their early training, in the same way that politicians of Abraham Lincoln's generation all had to learn how to perform the unnatural act of speaking from lecterns for hours on end in stentorian tones and flowery phrases.[8]

That is, the theory that politics today is corrupted by technology-based gimmickry is open to the facile retort that the solution is simply a matter of mastering the new techniques—not just by politicians but by the public.

DOMINATION OF INFORMATION FLOW

This suggests that we look for a less superficial theory, for a more probing critique of politics in an age dominated by the media. And such theorizing is not hard to come by. What I have in mind here is critics who worry that the media, broadly construed, (a) represent the primary source of information for most people today, and (b) are today increasingly falling into the hands of a smaller and smaller collection of idea managers, most of whom share a single political ideology. Conservatives, for instance, worry that the media are dominated by liberal reporters, whereas liberals and radicals both worry about corporate managers either dictating what will be reported, written about, or dramatized, or else hiring or retaining only writers who will (consciously or unconsciously) say what is favorable to the so-called "corporate liberal" *status quo*. I will not dwell on them here (see chapter 14, below), but conservative complaints about the allegedly liberal media have been around at least since Spiro Agnew's famous diatribe against "nattering nabobs of negativism." This point of view is best represented today by the group called Accuracy in Media.

The best known and perhaps still the best documentation of the influence of the media on electoral politics, and of the increasing transformation of the media into corporate entities, is David Halberstam's *The Powers That Be*.[9] That book is a journalistic compendium on fifty years of transformations of CBS, *Time*, the *Los Angeles Times*, the *Washington Post*, and the *New York Times*—but with special emphasis on CBS. Along the way, other newspapers, news and other magazines, broadcast companies and TV networks are discussed, along with advertising agencies, publishing empires, film production companies, government regulators, and so on. Everywhere the emphasis is on the individual people who made and make up these companies, beginning with the founding giants and ending with armies of network and publishing executives and their bottom-line-driven underlings. (Actually, Halberstam begins and ends with one of the giants, William S. Paley of CBS, but only to dramatize the change.)

The gist of Halberstam's message can be summarized fairly easily in the following paragraph (already dated after a decade of ever faster changes of the same sort):

> Time Inc. was changing significantly and quickly. Yes, it had added cable-television companies and bought the Book-of-the-Month Club, and yes, it was making some tentative moves into television, but it was not the same company. The dog had become the tail and the tail had become the dog. It was no longer a communications business with a resource ancillary, it was more and more a forests-products company or a resource company that also had some magazines. It was, with the acquisition of Inland, a $2 billion company, roughly the 150th-largest company in the country. Time executives estimated that with Inland the traditional Time Inc. editorial core operations—magazines and books—would now account for only about 45 percent of revenue.[10]

Some of this same ground is covered in Herbert Gans's *Deciding What's News: A Study of CBS Evening News, NBC Nightly News, Newsweek, and Time* (published in 1979, the same year as Halberstam's book).[11] Interestingly, Gans has explicit comments to make on the similarity between his sociological approach and the approach of a journalist (like Halberstam):

> In some ways, sociology and journalism are similar. They both report on American society through the use of empirical methods . . . [;] they face similar dilemmas in dealing with values, for both aim to be objective, even if neither can finally operate without values or escape value implications, however much the actual empirical work is value-free or, as journalists put it, detached.[12]

Gans, immediately after this, also notes the differences: "Being similar, sociology and journalism are, to some extent, in competition, so that not much love is lost between them." The sociological empirical method Gans uses is participant observation—though in this case that largely meant detailed interviews while watching people work. What interests me here is the values Gans turns up by this supposedly value-free inquiry (which might, coincidentally, have included Halberstam as one of his journalist subjects).

At the beginning of his study, Gans says that he discovered, in studying these journalists with a national audience, "how professional standards incorporate efficiency criteria and the realities of power" because these "journalists are, among other things, producers of symbolic consumer goods."[13] Later in the book, under the heading, "Objectivity, Values, and Ideology," Gans draws some explicit conclusions about values:

> Some of the people I observed were conservatives, a few were ultraconservatives, and a handful (not counting the house radicals of the 1960s) could be considered democratic socialists; however, the vast majority were independents or liberals.[14]

This would seem to support the charge of conservative critics of the media, but Gans immediately qualifies his claim:

> In the end, most of the people I studied could be classified as right-liberals [liberal on social issues, conservative on economic or social class issues] and left-conservatives [the reverse of the above].

More important, Gans adds that the journalists "are, on the whole, more liberal than their superiors and their colleagues in the business departments, as well as their sponsors and advertisers." He also notes that it is "the higher ranks who determine which stands will be taken on major issues."[15] (I should note that these are not just Gans's opinions; he claims that his interview data support these claims.)

Gans admits that his "empirical analysis looks at the status quo of the moment"—in this case, the mid-1970s—and might thus "overstate its permanence." This is particularly important to remember, looking back at Gans's conclusions after the Reagan decade. But Gans thinks little is likely to change under any administration, and his conclusions are interesting in any case. As mentioned, he lumps them under the headings of efficiency (the journalist's need to meet deadlines) and power (the tendency of journalists to report on powerful people, to depend on them as sources, and—he thinks to a very limited extent—to play an active role in bringing people to power). The conclusions are that journalists are leadership testers, suppliers of political feedback, power distributors (news makers as well as reporters), moral guardians (of mainstream values—and here Gans might have had something different to say in 1988 than in 1978), "prophets and priests" (here Gans reports what some commentators have said, apparently without much conviction that they are right), storytellers and myth makers, barometers of order (or barriers to the perpetually perceived threat of disorder), agents of social control, constructors of nation and society (it should be recalled that he is talking about journalists with a national audience), and "managers of the symbolic arena."

These final, supposedly empirically-based conclusions are items I will be returning to. But for the moment I want to return to Halberstam's analysis and a perspective that shows up only limitedly in Gans's analysis of the relationship between the media and political power. What I have in mind is the increasing "corporatization" of the media—something that continued to escalate in the decade after Gans and Halberstam published their books.

I cite just one source here, on a single issue (university presses and the mass-market publishing houses), but I do so because I think that issue hits home in a particularly clear way. Roy Reed, a journalism professor at the University of Arkansas, wrote an acerbic essay for the 1989 version of what is becoming a *New York Times Book Review* annual number devoted to university presses.[16] He there quotes the director of his university press, Miller Williams, to this effect: "I would almost say that the First Amendment guarantee is called into question if economic forces make it difficult to appeal to the larger society." The immediate presupposition of this cry of alarm about free speech and publishing is particular corporate decisions. Reed quotes Ben Bagdikian as reporting that "after the German conglomerate [Bertelsmann] bought Doubleday, it pruned the company's trade list by a third. That got rid of a lot of books that could not be spun off into movies, paperbacks or other profitable enterprises." The remote background—and the critics would say this is even more clearly a threat to free speech for writers—is the conglomerate move into publishing of which Bertelsmann is just one example. Reed again quotes Bagdikian as claiming that today only "five corporations dominate the world's communications industry," and he goes on:

> Three [of these] are especially interesting for the famous old names they now own. Time Warner, the newest conglomerate in the field, owns not only Warner mass-market paperbacks but also Little, Brown & Company, Time-Life Books and the Book-of-the-Month Club. A German company, Bertelsmann, owns Doubleday, Bantam Books, Dell and the Literary Guild. Rupert Murdoch's News Corporation owns Harper & Row and a big piece of Viking Penguin.[17]

And of course at this point we should recall Halberstam on Time Inc. ten years earlier, to the effect that publishing represented less than half of its diversified corporate interests. Reed was drawing an overly sharp contrast in order to dramatize the contributions that university presses make toward keeping our culture diversified today (and subsequent letters to the editor called him to task for his oversimplifications), but in the present context his point takes on chilling significance with respect to the possibility of dissenting activists getting their message across to enough people to bring about social change. And the issue is not so much ideological control of what people can say and where they can say it as it is a matter of the clichéd "bottom line" severely restricting options to what will be commercially successful in selling to a mass audience. This sounds much more somber than Gans's reassurances about national journalists still serving as "symbolic managers" for society. Indeed, it suggests that we might well worry about Gans's other conclusion, that these people (now seen as severely constrained by demands to remember the bottom line) also serve as agents of social control. That is, it is possible to interpret Gans's liberal-values-oriented data as suggesting a much more radical interpretation.

ECONOMIC CONTROL OF THE MEDIA

The most articulate—some would say shrill—critic of the corporations as consciously striving to control public thinking through control of the media is Herbert Schiller, in a series of books from *Mass Communications and American Empire* to *Culture, Inc.*[18] The subtitle of the latest book, *The Corporate Takeover of Public Expression*, aptly summarizes the thrust of Schiller's critique.

On the first page of the introduction to *Culture, Inc.*, Schiller spells out, very clearly, his basic orientation. He says he is worried primarily about the "consolidation of corporate power" in the United States, and that at the expense of "once important forces in American life—independent farmers, organized labor, and a strong urban consciousness." At that point, all he is willing to say is that, "This imbalance has much to do with the changed role of information in the economy."[19] By the end of the introduction, this tentative beginning has been transformed into a strong hypothesis to be tested throughout the remainder of the volume.

Schiller's thesis is that for most of the twentieth century, but increasingly in the period following World War II, there has been a steady shift of income from the poor and middle class to the richest segment of American society, *but* this fact has been rendered invisible by an amazingly successful public relations coup on the part of corporations. Schiller first indicts both Democratic and Republican administrations: "This 'invisible' income redistribution is the [direct] contribution of the postwar national administrations." But they did what they did for a reason: "It is . . . an indirect and generally unpublicized testimonial to the successful informational efforts and energies expended by the corporate sector and its media champions." This leads to Schiller's basic question:

> How to account in an age of intensive publicity for such reticence about such a remarkable public relations achievement—the transfer of wealth from the poor to the rich without a sign of public indignation?

And here is Schiller's proposed answer: "This 'achievement' may be regarded as one of the tangible outcomes of the corporate envelopment of public expression." Further, "A precondition for this 'achievement' has been the growth of the information-cultural industries where the messages and images that influence public consciousness are produced."[20]

A run-through of the chapters in *Culture, Inc.* is a good way to summarize the evidence Schiller amasses in support of his thesis. Chapter 2, "The Corporation and the Production of Culture," rehearses (and enlarges on) the evidence already alluded to here of the consolidation of the whole range of the media—from TV, including cable, to newspapers and newsmagazines, to the entertainment industry, to book publishing and national bookstore chains, and on and on—in the hands of an ever smaller number of corporate heads. Schiller does raise an objection to the sweeping nature of his indictment here, asking whether there are not some creative individuals who buck the corporate commercial trend. But he thinks, sadly, that the evidence supports rather than challenges his thesis:

> Even those who may appear to be the last holdouts of individual creativity, the studio artists working alone, find no escape from the market imperative, though the pressure may be brought to bear in less explicit ways. The gallery system, private collectors, art speculators, and the process of museum acquisition constitute a special but in no way fundamentally different commercial framework than television networks that [similarly] commission shows from [supposedly independent] TV production companies.[21]

Chapter 3, "The Corporation and the Law," shows how the courts, including the Supreme Court, have put a cloak of legality around the right to engage in commercial speech as a protected freedom alongside all other forms of free speech—where, Schiller would say, commercial free speech tends to drown out all the other kinds. Again Schiller raises an objection to his view, but the only real challenge to commercialized free speech that he can find is dangerous invocations of national security, in the Reagan years, that might subordinate the now-legally-secured right of commercial free speech to the demands of either "law and order" internally or national security externally.

Chapter 4, "Privatization and Commercialization of the Public Sector: Information and Education," spends a great deal of time documenting how the information industry—computerized libraries and a whole range of related services—were either privatized or directly linked to corporations. (There is here a passing reference—just two pages—on commercialization of higher education in joint business-university ventures.) Chapter 5, "The Corporate Capture of the Sites of Public Expression," focuses on the commercialization of museums, the "malling" of America at the expense of cities as "public spaces," and corporate control of what many still think of as public expression, public television. Finally, chapter 6, "The Transnationalization of Corporate Expression," shows how all these trends are then extended overseas in the commercialization-through-control-of-information of the whole world.

Schiller, once again, is open to objections to his view, and chapter 7, "Thinking about Media Power: Who Holds It? A Changing View," is devoted in large part to dealing with communications scholars who were downplaying the effects of the media at the very same time the corporations were taking over their control, both nationally and internationally.

In the end, Schiller really cannot understand the blindness of the communications scholars, and he concludes: "In the late 1980s, the control of representation and [symbolic] definition remains concentrated in the products and services of media-cultural combines"—and Schiller thinks it is obvious that there is a reason for this state of affairs. His answer? "That control can be challenged only by political means."[22] He is not overly optimistic on this score: "Postwar developments in the informational-cultural sector at home and abroad do not give much encouragement to expectations for an expanded public sector of communication and cultural expression."[23]

In this final conclusion, Schiller joins his fellow radical critic of late capitalism, Herbert Marcuse, with his thesis about "one-dimensional man." The managerial classes in advanced capitalist societies (and, Marcuse added long before the end of the Cold War, in the managerial societies of the Eastern Bloc as well) have the power, through judicious use of the educational system and the mass media, to render dissent extremely unlikely if not impossible. This radical version of the media-control thesis I will reserve for chapter 11. For now, I want to return to less sweeping theories—and to the possibility of activism to control the media.

PROGRESSIVE ACTIVISM AND THE MEDIA CONTROL OF POLITICS

It is by no means necessary for one to agree with these radical indictments of corporate control through manipulation of the media in order to be concerned about the negative impact of the media on democracy. Progressive liberals too have been concerned over the capture of political institutions by corporate interests. I focus here on two organizations, John Gardner's Common Cause—the preeminent "citizens lobby" in the United States today—and Ralph Nader's consumer group, Public Citizen.

Shortly after Gardner founded Common Cause in 1970, he wrote a book both to explain and to justify the venture; the book was entitled, *In Common Cause*.[24] It was not intended to be either a sociological or a journalistic account, but it did attempt to summarize some of the data already referred to here about the capture of politics by monied interests. Neither did it pick out corporations in particular as the culprits. What Gardner focused on, instead, is the by now clichéd "unholy trinity" of interests that control government in the interest of the few and at the expense of the democratic majority:

> One of the realities that constantly undermines the public interest is the existence, at both federal and state levels, of what might appropriately be called the unholy trinity. It takes many forms, but most typically it consists of an upper-middle-level bureaucrat (say, a bureau chief from one of the cabinet departments), a legislator (say, a member of one of the appropriations committees), and a lobbyist from one of the well-heeled special-interest groups. As a rule, they will have been friends for years. . . . They have seen Presidents, cabinet members, and governors come and go. . . . They are part of the permanent, invisible government. And as the years go by they shape policy in substantial ways.[25]

And of course Gardner would have us understand that they shape policy, not in the interest of the general public (and certainly not in the interest of those at the lower end of the socioeconomic scale), but in the interest of those already in power—including themselves. This was the sort of cabal that Common Cause went after in its early years, attempting to limit the powers of committee chairs in Congress, to open up committee hearings (including budget writing and appropriations), and joining in the fight to disclose the Watergate scandal and obviate that sort of scandal in the future. Archibald Cox, one of the heroes of Watergate, followed in Gardner's footsteps as chairman of Common Cause for about a decade.

This focus on the corruption of money in politics has continued over the two decades of Common Cause's existence. In the late 1980s and early 1990s, the focus is on the control of political action committees. For example, *Common Cause Magazine*, July/August 1990, includes a centerfold extra on the latest news from Congress: "Common Cause wins dramatic campaign finance and honoraria victories in Senate." But it had to add, in smaller type: "House passes flawed campaign reform bill." After more than five years of work by thousands of volunteers all over the country, this was the best that could be reported at the time. And of course skeptical critics have been quick to note that the PAC scandals were in large part an effect of legislation passed in the wake of Watergate that encouraged the growth of PACs. (This sort of irony has never fazed Gardner; his view, often stated, is that reform efforts are never final—and reforms always call for new reforms as politicians learn to get around the old ones.)

There is another sort of irony in recent Common Cause proposals for campaign reform. Gardner had, in *In Common Cause*, called for citizen activists to make full use of the media: "The light of day," he said, "has a marvelously cleansing effect on politicians."[26] For them to be able to do so effectively, the media would have to be less captive than Schiller says. And a great many people have commented that one of the reasons there is so much PAC money in today's campaigns is the costs of TV air time, not to mention the costs of all the technogimmickry of polling, focus groups, etc., mentioned earlier. In recent years, Common Cause's leaders have certainly not been unaware of these claims about media corruption of campaigning, but they have so far focused mainly on the control of PAC spending and public funding of congressional campaigns—presumably including public funds for both TV time and technogimmicks.

Nader's group, Public Citizen, has from the beginning had a focus different from that of Common Cause. One difference is obvious, that between Nader himself and Gardner. Though both Common Cause and Public Citizen (including its host of satellite or spinoff organizations) have developed a second generation of leaders, Nader has remained in the public eye and continued to spearhead the crusades of his organization(s) much longer than Gardner. He is also much more of an outsider.

As a gadfly for American democracy, Nader has few rivals, now or throughout the history of the country. From the beginning, Nader has excelled at playing the media to his (and, he would say, to consumers') advantage. This has included frequent press releases and press conferences on key issues. One early assessment of public interest lobbying, by Jeffrey Berry, notes: "The fact that Nader will appear at a press conference makes the conference a recognized media event."[27] Berry goes on:

> The third way in which groups may use the media directly to propagate research and other material is the cultivation of individual reporters. . . . [This] tactic of conducting and then publicizing research has been used to the greatest extent by Ralph Nader and his many associates.

Berry's leading example is *The Nader Report on the Federal Trade Commission* published in 1969, but he also refers to studies of the Volkswagen, of DuPont influence in Delaware, of old age homes, and of occupational health and safety, among many others.[28]

Berry also notes some ways in which Nader's Public Citizen is similar to Common Cause. He refers to another study, *Who Runs Congress?* by Mark Green and others, then he says:

> In the past few years, Nader has tried to create a "second strike" capability to back up his numerous study groups. The first move in this direction was the establishment of the Public Citizen Litigation Group. . . . Another effort by Nader was his founding of Congress Watch, a lobbying group that is concentrating on consumer affairs issues.[29]

A decade later, another student of progressive public interest activism, Michael McCann, could still pick Nader out of the crowd as perhaps the leading example:

> Like their radical New Left precursors in the 1960s, public interest activists have professed to take participatory democracy quite seriously as a basic commitment of the movement. "We strive for a reawakening of the democratic impulse—the promise that people can shape the decisions which affect their lives," proclaims [*Public Citizen 1979 Annual Report*].[30]

And McCann goes on to cite some Nader associates as saying: "Power to the people does not mean a few spokesmen leading around a disinterested flock of 200 million consumers"; it means getting the people involved in "letter writing, telephone calling, speech making, pamphlet distribution, demonstrations, teach-ins, and various forms of direct lobbying."[31] For this purpose, numerous how-to manuals have been produced, including Donald Ross's *A Public Citizen's Action Manual*.[32]

In the context of this chapter, there is more than a little irony in the relationship between these public interest groups and the media. With very few exceptions—and the most notable exception is the Media Access Project—the public interest groups have not focused on cries of alarm, such as those by Schiller, about corporate control of the media. They seem to assume that, because Nader and his groups and Common Cause and environmentalist groups find it relatively easy to get results of their studies in the press and on television, the media are an ally of progressive, populist democracy. They do not even seem overly concerned—if we can judge, for instance, from the relatively low priority the issue has so far had for Common Cause, arguably the most effective of the public interest groups—about the negative influences on electoral politics of TV campaigns, focus groups, electronic targeting of particular audiences, etc., of all those things I have here labeled "technogimmickry."

Some groups, such as the Media Access Project and People for the American Way, do address these issues. MAP, for instance, from the beginning made this its slogan: "An informed electorate has [always] been essential to the working of democracy."[33] And one MAP study attempted to demonstrate a decisive link between corporate spending on popular referenda and their outcomes.[34] But for the most part the public interest groups seem to assume that the media can be used effectively to win the public over to their side.

To understand this, we might best return to the study by Herbert Gans. As noted, he believes that his data show the vast majority of journalists addressing a national audience (on TV and in the newsmagazines) are centrists—either at the left end of the conservative spectrum or (more commonly) at the right end of the liberal spectrum. They can, Gans says, easily be swayed toward the left on social issues—

the focus of the public interest groups, for the most part—even though they tend to lean more toward the right on economic and social class issues. And of course their bosses tend even more toward the right.

In spite of his evenhandedness on the issues addressed in this chapter, Gans does think activism is called for on some issues. His call for reform, which he calls "multiperspectivism," while quite radical in some senses, is very non-radical in the usual political sense of that term. What Gans wants to change is the personnel of the national news media—the professionals in the field. He first states his goal:

> I would argue that the primary purpose of the news derives from the journalists' functions as constructors of nation and society, and as managers of the symbolic arena. The most important purpose of the news, therefore, is to provide the symbolic arena, and the citizenry, with comprehensive and representative images. . . . In order to be comprehensive, the news must report nation and society in terms of all known perspectives; in order to be representative, it must enable all sectors of nation and society to place their actors and activities—and messages—in the symbolic arena.[35]

Gans recognizes that this is utopian, and what he really wants is just more perspectives to be represented than there are currently—especially perspectives of the poor, ethnic groups, etc. He wants government to support this venture (as it does in several European countries), but he also worries about the possibility that would bring of government control and his proposal calls for an independent agency called an Endowment for News similar to the U.S. national endowments for the arts and humanities. He recognizes that this is a political matter—undoubtedly, if he were making this proposal today, he would recognize how political—but he thinks the advantage would not fall to any particular party but to democracy itself. In his proposal, Gans says, "The symbolic arena would become more democratic" and "the symbolic power of now dominant sources and perspectives would be reduced."

How feasible is such a proposal? This is a question Gans asks himself in his concluding paragraphs. His answer is pessimistic; there would, he says, be little "incentive for multiperspectivism because it undercuts the twin bases of the present journalistic enterprise: efficiency [in meeting deadlines] and the power of sources." That is, we ought not to expect much from the professionals in journalism, broadcast or print. (And if we broadened Gans's ideas to cover other aspects of the media, publishers and editors, actors, etc., we ought not expect too much from them either.)

Nonetheless, Gans remains optimistic, and I wish I could too. It seems true that professionals in the media today experience enormous pressure, generated by corporate owners and sponsors, to respond to the demands of the clichéd bottom line. (Jacques Ellul, in the second quote I included at the head of this chapter, draws these implications out to their absurdist conclusions.[36]) But this very fact suggests that these same actors, supposedly intelligent and creative people, *could* do things differently if they recognized widespread demands for social responsibility on their part.

So here my conclusion is about as pessimistic as any I will arrive at in this book. Major changes seem to be needed in the way the media operate today, and in particular in their negative influence on the democratic process. Yet there are only a handful of activists, whether within the ranks of media professionals themselves or in public interest activist groups, working to undo this damage.[37] The only optimistic thing I can think of to say is that if enough people become aware of the enormity of the problem, we *might* hope for some increase in the activism that seems to be called for.

Part III

Specific Technosocial Issues

7

Biotechnologists, Scientific Activism, and Philosophical Bridges

In the end, the security we fought so hard to preserve will have disappeared forever. Thanks to bioengineering, . . . we will no longer control any measure of our own destiny. Our future will be determined at conception. It will be programmed into our biological blueprint.

—*Jeremy Rifkin*

Having considered a set of general social problems that vex our technological society, I turn now to problems more directly related to the scientific/technological community. Over the last two decades, one of the most hotly contested issues of this sort has been biotechnology. In this chapter, I concentrate on just one feature of that public debate involving scientists—namely, the debate over recombinant DNA in the 1970s. I attempt to put this within the larger context of earlier scientific controversies. And I end the chapter by relating this scientific activism to the philosophical approach I am defending in this book.

In the public controversy over recombinant DNA in the late 1970s and in much public discussion since, fears of bioengineering or biotechnology are often expressed in the most general of terms.[1] Since I do not think that reasonable discussion is fostered in that fashion, I want to focus my discussion here on specific fears that have been expressed about specific biotechnology developments.

Neil A. Holtzman, in *Proceed with Caution: Predicting Genetic Risks in the Recombinant DNA Era*, says that, "There is no retreating from the use of genetic tests," and his evidence for that claim seems to me persuasive. What his book is about, then, is not stopping such uses but controlling them in a way that respects our democratic civil liberties. He ends with a list of the controllable dangers that he has discussed:

> Just as I deplore the imposition of testing on anyone—or the imposition of any action pursuant to a positive test result—so I deplore withholding these technologies from those who would freely and knowledgeably choose to use them. The major dangers of genetic testing in the recombinant DNA era are its unbridled development without assurance of test validity and reliability;

> inadequate knowledge on the part of providers and potential recipients; inequities in test use and ensuing care; and, finally, restrictions on individual autonomy and confidentiality.[2]

Holtzman not only details these dangers but proposes specific policies that might help to minimize them. This seems to me an admirable thing for a biomedical researcher to do.

The American Association for the Advancement of Science and its Committee on Scientific Freedom and Responsibility have also focused on specific problems associated with specific biotechnology developments. Mark S. Frankel, in his preface to *Biotechnology: Professional Issues and Social Concerns*, summarizes two issues that have received much discussion among socially responsible researchers. One concern is linked to the broad area of risk, which so troubled the activist critics in the recombinant DNA controversies, as well as to efforts to contain the risks:

> These differences among scientists and the level of public concern have had several effects. Some experiments have been substantially delayed as uncertainty about risks has led to public protests and prolonged litigation. In this country, regulation of biotechnology has proceeded spasmodically at the federal level and still creates confusion among researchers. A "patchwork pattern of local laws" may be emerging at the local level which will only add to the confusion. And since microorganisms do not recognize national boundaries, there have been calls for international guidelines.[3]

Frankel concludes the passage this way: "The controversy over the potential hazards of genetically engineered organisms is unlikely to dissipate in the near future."

The specificity in this first case has to do with efforts by governments to contain risks; the second area of concern addressed by Frankel and the AAAS volume he is introducing has to do with corporate-academic links established to promote the development of biotechnology. Here is Frankel's summary of the concerns under this heading:

> Such collaboration has . . . prompted concerns about its impact on the values and mission of universities. These concerns, in one form or another, have encompassed issues of academic freedom, conflicts of interest, the university's public service role, proprietary information and publication, the research priorities of faculty, and possible constraints on the education of students.[4]

Still other concrete concerns have been raised, in the journal, *Agriculture and Human Values*,[5] about uses of biotechnology in agriculture.

Gary Comstock, in "The Case against bGH," argues that there are two fundamental moral arguments against the attempt to engineer, using bovine growth hormone, cows that will produce more milk: "its treatment of animals, and its dislocating effects on farmers." His conclusion emphasizes the latter argument:

> To the extent that potentially displaced dairy farmers have done nothing for which they ought to be punished; to the extent that the research establishment has clearly favored large producers in its development of techniques and technologies; to the extent that fiscal, monetary and economic policies have disadvantaged small dairy producers; and to the extent that [bovine growth hormone] will only exacerbate the unjust consequences of the past; to that extent we ought to oppose this particular biotechnology.[6]

Mark Sagoff, in "Biotechnology and the Environment: What Is at Risk?," plays off against one another two views of what is at stake in setting environmental policy with respect to biotechnology. He first cites "an extremely influential view among economists" that "holds that a regulatory agency such as [the Environmental Protection Agency] should be concerned primarily or, indeed, exclusively with improving the overall efficiency of the economy." This, he says, is consistent with promoting biotechnology almost without limit. On the other hand, he asks: "What would the regulatory and research agenda be like if EPA, instead, conceived of itself as an *environmental* protection agency?" (Underlining is Sagoff's.) He answers: "Its task would then include protecting the evolutionary and ecological *status quo*, even if that meant keeping economically beneficial rDNA organisms out of the natural environment." Clearly Sagoff prefers this policy option, and he says it is clearly "in line with [EPA's] statutory authorities, which require it to protect the nation's ecological and evolutionary heritage."[7]

Even if one does not accept Sagoff's choice between two policy alternatives, his article serves to suggest that one important concrete issue in biotechnology is the correct trade-off or balance along a spectrum between nature and technology in this area of development.

J. Sousa Silva, in "The Contradictions of the Biorevolution for the Development of Agriculture in the Third World: Biotechnology and Capitalist Interest," claims that, "The assumption that biotechnology will solve Third World social and economic problems through an agricultural revolution is a myth."[8] More specifically, he maintains that it is a myth devised to support the continuation of capital accumulation, primarily in the countries in which biotechnology is developed. Again we do not have to accept the either-or formulation—it will be beneficial either for Third World countries or for first-world capitalists—to acknowledge that very real and very concrete questions about whose interests are being served are at issue when biotechnology is exported to Third World countries.

Unlike vague fears about releases of genetically altered microbes that initially bothered the critics of recombinant DNA research, these are quite specific and detailed criticisms of particular problematic aspects of biotechnological development. In this chapter, I address ways in which these sorts of problems—about questionable uses of genetic testing, potentially excessive and confusing government regulations, possible conflicts of interest, possible injustices against small farmers, environmental trade-offs, and dependency issues for Third World countries—can be addressed in a reasonable fashion both by academics and by socially responsible members of the technical community outside academia.

THE ALARMIST VIEW

Jeremy Rifkin is widely recognized in the United States as the preeminent critic of genetic engineering. In his book, *Algeny*, he has summarized both ethical and political concerns. "Proponents of human genetic engineering contend that it would be irresponsible not to use this powerful new technology to eliminate serious 'genetic disorders.'" Citing concerns of *New York Times* editors and of bioethicist Daniel Callahan, Rifkin retorts: "Genetic engineering poses the most fundamental of questions. Is guaranteeing our health worth trading away our humanity?" He goes on to state this objection in political terms—"All biologically engineered products require that someone make a decision about which genes should be engineered and which genes should be done away with"—and he concludes with a pessimism worthy of Aldous Huxley a generation earlier:"Thanks to bioengineering, . . . we will no longer control any measure of our own destiny. Our future will be determined at conception. It will be programmed into our biological blueprint."[9]

Rifkin is not alone in his sense of alarm. Philosopher Hans Jonas shares Rifkin's pessimism, and he makes it an issue of the deepest philosophical significance. Like Rifkin, Jonas has been recognized for a long time as an opponent of genetic engineering, opposing it as part of an "ethics of caution"—granting the awesome power of recent technologies such as bioengineering, humans absolutely *must* consult their fears rather than their hopes in setting moral priorities. Here is one particularly eloquent (albeit sexist) formulation:

> The idea of making over man is no longer fantastic, nor interdicted by an inviolable taboo. If and when that revolution occurs, if technological power is really going to tinker with the elemental keys on which life will have to play its melody in generations of men to come, . . . then a reflection on what is humanly desirable and what should determine the choice—a reflection, in short, on the image of man, becomes an imperative more urgent than any ever inflicted on the understanding of mortal man.[10]

Later in the same essay, Jonas says, "One part of the ethics of technology is precisely to guard the space in which any ethics can operate"—where the same assumption is being made by Jonas as had been made by Rifkin, namely, that a genetically determined life leaves no space for ethical choice.

In general, scientists involved in genetic engineering have paid little heed to such alarmist fears; instead, they have focused on the more manageable problems I listed above.[11]

SCIENTISTS RESPOND

Sheldon Krimsky, in *Genetic Alchemy*, summarizes how scientists and other science-related professionals in academia have gotten involved in sociotechnical controversies during the forty-plus years since the Second World War. Krimsky's focus is the public debate over recombinant DNA research in the 1970s, but he prefaces that with an account of scientists' involvement in earlier controversies that paved the way for their involvement in the rDNA debate. As he says, "Much of the post-1973 political development is clearly visible in these precursor cases."[12]

Krimsky begins with the Federation of American Scientists, which he calls "the oldest, largest, and strongest" of the groups that would eventually be involved in the rDNA debate. The roots of the FAS were in the Federation of Atomic Scientists, which had, after World War II, worked to internationalize control over nuclear power and especially nuclear weapons. But its most immediate battle, before the rDNA debate, was a late-sixties fight over deployment of a U.S. anti-ballistic missile system. FAS has always been a politically moderate group, and Krimsky notes how resolutely it opposed Communist influences in its ranks from the very beginning; in this he contrasts it with a slightly older group, the American Association of Scientific Workers, about whom Krimsky says it is probable that it included a number of members of the Communist Party USA.

"Several other groups," Krimsky notes, "were similar to the FAS in their conception of the 'special responsibility' of scientists to inform and, if need be, instruct the public on matters of technoscientific importance." He cites the Society for Social Responsibility in Science, which, beginning in 1949, worked on constructive alternatives to militarism; the Scientists' Institute for Public Information, which concentrated on nuclear testing issues in a broader context of the growing environmental movement; and finally on scientists who banded together with other moderate activist groups to oppose the building of an American supersonic transport (SST).

Krimsky contrasts these moderate scientific groups with more militant, radical groups of scientists. In addition to AASW, already mentioned, he lists the Medical Committee for Human Rights, originally organized to provide a medical presence in the civil rights movement of the 1960s but later broadened to bring radical pressure to bear on various aspects of the capitalist health care system in the United States; the New University Conference, an outspoken radical group of faculty and graduate students involved in battles over military training on campus but best known for opposition to the Vietnam War; and, mentioned in greatest detail, Scientists and Engineers for Social and Political Action, whose offshoot, Science for the People, played a key role in the rDNA debate but had earlier been involved in controversies over sociobiology, IQ testing, and genetic research generally.

Krimsky summarizes the activist work of the moderate scientific advocates of social responsibility this way:

> For the moderates, the most feared impacts of science on society were the unanticipated or unintended ones, and the responsibility of the scientist consisted in calling the attention of the public to these dangers. . . . Indeed, in the rDNA debate, the concerns of this group centered on the problem of laboratory accidents . . . [and] proposed solutions took for granted both the ability and the good faith of the scientific community to neutralize these possibilities.[13]

On the other hand,

> The radicals . . . looked at the impact of science on society principally in terms of how science served to strengthen existing economic and social relations. . . . The aspect of gene-splicing technology on which they focused initially was the potential it held for genetic engineering, and this was not something that could be dealt with by a simple technological fix.[14]

There is then, Krimsky is suggesting, a range of scientific/technological activism. At one extreme are scientists and engineers who believe in the objectivity of their work and fight against alarmism by attempting to educate the public. At the opposite extreme are radicalized scientists out to eliminate the negative influences of capitalism on science and technology, making them serve the interests of the broader public rather than moneyed interests. Although not mentioned by Krimsky in these passages, there is a broad spectrum of scientific and other technical activists between these extremes—not to mention a much larger percentage of scientists, engineers, medical researchers, etc., who believe activism is not part of their responsibility.

PHILOSOPHICAL PARALLELS

I want at this point to link my concerns with scientific activism (not only in this chapter but in the three to follow) with the philosophical approach espoused throughout the book.

For well over a generation, a narrow academicism—similar to the objectivity claims of scientists and the neutral-technology claims of engineers—has held American philosophers in thrall. Now a new movement is afoot, and many philosophers are rebelling against academicism.[15] They want to build bridges to segments of society outside academia, in government, industry, and groups representing the public. In many ways this represents an old, almost lost tradition in American philosophy. Bruce Kuklick, chronicler of the period in which the older tradition was lost, begins a crucial chapter on William James and Josiah Royce this way:

> Throughout the first decade of the [twentieth] century each developed his moral and religious philosophy and applied it to practical problems, communicating the results to audiences outside the academy. . . . [B]y using theory to understand man's place in the world . . . the philosopher had a public duty to speculate on real-life problems.

This was before the First World War; later, Kuklick notes:

> By the Second World War specialization within the profession had changed the substance of speculation and radically reoriented the traditional meaning of philosophy. It lost its place as the synoptic integrator of the manifold intellectual concerns of human existence.[16]

The loss of relevance was neither immediate nor universal. Until his death in 1931, George Herbert Mead (one of the models for my approach; see chapter 2, above) remained a committed philosopher-activist in and around the University of Chicago. According to Andrew Reck:

> Education did not hold a monopoly on Mead's readiness to contribute to social philosophy and practical affairs. Social justice and governmental reforms were also close to his heart. He wrote articles on such topics as settlement houses, criminal justice, philanthropy, . . . social security . . . [and] the motion picture as a form of artistic expression.[17]

Mead also worked with activist groups in and around Chicago on almost all of these issues.

John Dewey (as also reported in chapter 2) carried this philosophical activism to a national scene and to heights of activism that far surpassed Mead's. Gary Bullert, in *The Politics of John Dewey*, summarizes Dewey's activities:

> Dewey's public career existed in an age of radical political upheavals and cultural dislocations. He underwent the trauma of World War I, the Scopes Trial, the Sacco and Vanzetti Trial, the Great Depression, the rise of totalitarianism, the Trotsky Trial, World War II, and the Korean War. Modern mass society's erosion of traditional beliefs and social bonds jeopardized both political freedom and cultural stability. . . . Becoming America's preeminent public philosopher, Dewey tirelessly addressed himself to the public issues of his day.[18]

Moreover, Dewey articulated even better than Mead the ideal of a publicly committed academe. Again according to Bullert, Dewey maintained that, "Scientists and scholars inherited the social responsibility to propagate the scientific attitude. . . . By practicing . . . openmindedness and rationality, scholars could elevate the level of political discourse for the entire community." Bullert summarizes Dewey's view this way:

> How could academics reconcile intellectual and political responsibility? This issue is crucial in Dewey's career. Insulation from political power could ferment alienation and political irresponsibility. Total immersion in government and competitive policies might undermine objectivity. A healthy democracy requires a sensitized equilibrium between scholars as a cloistered leisure class and servants of society.[19]

Today, the burden of building philosophical bridges to the world outside academia has fallen on people calling themselves "applied ethicists," some of whom have even ventured opinions on the ethics of bioengineering.[20]

I take up my quarrels with applied ethics later in the book, but I need to anticipate here. In my opinion, the best treatment of applied ethics that focuses on scientists, engineers, physicians, nurses, lawyers, accountants, etc., is that of Michael Bayles in his *Professional Ethics*.[21] He places professional norms midway between concrete cases and a set of common democratic values, assuming that some sort of valid deduction will allow us to arrive at a correct solution of particular problems based on these value premises. However, he is astute enough to recognize that assuring compliance with ethical norms requires something more. It may be more laypersons on regulatory boards, more effective reporting procedures, more effective enforcement, preventive measures, including better ethics education during professional training or an "ethical climate" in the workplace—and, if all else fails, government regulation or lawsuits.

An anthology on professional ethics edited by Joan Callahan, *Ethical Issues in Professional Life*,[22] better typifies the standard approach to applied ethics as dependent on standard academic ethical theory. Referring to Richard Brandt, Norman Daniels, Ronald Dworkin, John Rawls, and Stephen Stich, among others, Callahan describes a common approach called "reflective equilibrium." This involves proposing a set of moral principles to be compared with basic moral intuitions about appropriate behavior in particular problematic situations. Sometimes these intuitions will need to be brought in line with moral principles, but at other times the moral principles will need to be adjusted to accommodate real-life applications. In the hands of most applied ethicists, this approach leads to subsuming particular cases under general moral rules—even when "reflective equilibrium" requires reformulation of the rules.

I believe that this approach can lead to problems when philosophers attempt to build bridges to the people who must make the tough decisions in problematic cases—for instance, scientists and others involved in biotechnology decisions. One problem shows up if the philosopher takes the role of a "moral expert," telling the decision makers what the right moral decision is. In too many instances, the practitioner may grumble, asking what the philosopher knows about research or medical care or the real-life obstacles to implementing the allegedly "right" moral decision. But even if those faced with the practical decisions defer to moral experts, a problem remains. Simply implementing correct decisions supplied by an expert does not necessarily mean the decision maker will have learned anything about how to deal with the next problematic situation that arises.

There is another sort of difficulty if the theoretical philosopher takes on another traditional role, that of the "wise man" (or woman). I would be the last person in the world to say that we do not need wisdom in solving the practical problems associated with biotechnology—or any other technology. But wisdom, like "moral expertise," tends to strike practitioners as pontificating from on high, with the same two problems associated with moral expertise—too easy deference, or practitioner-vs.-theorist resistance.

What Dewey would suggest, and I think this is the right way to build bridges, is to pitch in with the groups trying to work out a solution democratically. In the examples cited at the beginning, this could mean something like the following:

—working with civil libertarians (who, in their turn, would do best to work collaboratively with the genetics researchers, insurance executives, etc.) to make sure that genetic testing is not used in inequitable or discriminatory ways;

—working together with genetics researchers and government regulators to achieve both balanced regulations and evenhanded enforcement;

—working with university committees to set appropriate limits on academic biotechnology work done for profit or on corporate-academic contracts or other joint ventures (which, again, would almost certainly also involve the philosopher in contacts with corporate spokespersons);

—working with both advocates and critics of agricultural biotechnology to help them decide what is fair in clashes between, say, large agribusiness enterprises (often supported by agricultural researchers in academia) and small farmers claiming they are being put at an unfair disadvantage;

—working with both EPA regulators and biotechnology advocates to achieve a balance between ecology and development; and, finally,

—working with advocates of both "appropriate technology" and "technology transfer" in wrestling with the extremely thorny issues associated with applying biotechnology (and other technologies) in the Third World.

Here we should remind ourselves of Dewey's ideal, cited

earlier—namely, that scientists and scholars have a duty to "practice openmindedness and rationality" in such ventures, and to help others do the same. In many cases, all the contemporary academic philosopher has to offer to others is his or her training in clear thinking, in how to formulate questions clearly so that reasonable answers are possible, in translating loaded ideological language into a neutral idiom that will help to defuse animosity in discussion, and so forth. These are skills that almost every philosopher, of whatever school or tradition, is expected to learn today. And practitioners often seem to appreciate this sort of issues clarification. (In my experience, they are as positive in their appreciation of this contribution as they are negative in their rejections of "moral expertise" or alleged "wisdom" on the part of practitioners of "applied philosophy.")

This role of philosophical bridge-building in the solution of social problems associated with biotechnology (or other technologies) does not preclude advocacy on the part of applied ethicists. They can, usually, easily separate their role as issue-clarifiers from their role as advocate of one viewpoint or another in whatever dispute it is in which they are intervening. They might, at that point, preface their remarks by saying, "I am speaking as a layperson on the issue"; but there is no reason why they cannot also make themselves as expert (*not* in the sense of "moral expertise") as other non-technical people often do in such controversies—or as technical people do when they speak out reasonably on an issue outside the area of their technical training.

This is a note to end on in this first of four chapters devoted to scientific/technological activism. Academicism in philosophy, along with the denial of social responsibility on the part of scientists (hiding behind a mask of objectivity) or engineers, including bioengineers (claiming "technological neutrality"), seems to me inappropriate, especially if the radical activists are right that technological ills are associated not with evil choices of individual decision makers but with social structures. And there are significant problems associated with developments in biotechnology—even apart from the "big" fears of some dangerous recombined microbe running loose in the general population, or of some anti-democratic geneticist engineering the "perfect human" in some "brave new world." Neither academics nor scientists and engineers in industry or government need to agree with the diagnoses of the radical activists. But it seems to me that if their work leads to significant social problems—or even a public perception of problems—they have a social responsibility to help look for solutions.

8

Computer Professionals and Civil Liberties

[There is] a set of ethical issues that surround computers emphasizing what the issues mean for computer professionals. Often the issues call for some sort of societal response—laws, regulatory policy, and so on. At other times, the issue cannot be dealt with fully at the societal level but, rather, requires responsible behavior on the part of . . . those who understand and work with computers.

—*Deborah Johnson*

Having considered the obligations of professionals, we face questions concerning how one ensures compliance with them. How does one prevent misconduct . . . from occurring? If misconduct occurs, what responses are appropriate? Most discussions focus on sanctions . . . assuming that adequate sanctions deter misconduct by others. That assumption is doubtful, at least at present levels of sanctioning.

. . . Even if there are good reasons against self-regulation . . . , one must consider whether alternatives [administrative agencies, lawsuits, civilian rather than professional review boards] are any better.

—*Michael Bayles*

Since social scientists first floated proposals for a national databank in the United States in the early 1960s, cries of alarm have been sounded over computer threats to individual privacy. Such cries have been heard ever since, though now they have been broadened from worries about databanks to concerns over government surveillance, corporate spying, and criminal eavesdropping affecting people not only at work but in the privacy of their homes. In a nutshell, many people worry about Big Brother watching (whoever the perpetrator may be thought to be).[1] This chapter focuses on both ethics[2] and civil liberties[3] aspects of a well defined set of issues.

THE PROBLEMS ADDRESSED HERE

The set of problems I am concerned with have been summarized by the Office of Technology Assessment (OTA), an arm of the U.S. Congress. One OTA study says:

New technologies—such as data transmission, electronic mail, cellular and cordless telephones, and miniature cameras—have outstripped the existing statutory framework for balancing . . . interests [in civil liberties and law enforcement].[4]

This October 1985 OTA report was the first of several under the general heading: *Federal Government Information Technology.* Its subtitle, *Electronic Surveillance and Civil Liberties* suggests something of the scope of the study. Chapter 1, a summary, tells us more:

—The number of Federal court-approved bugs and wiretaps in 1984 was the highest ever.
—About 25 percent of Federal agency components responding (35 out of 142) indicated some current and/or planned use of various electronic surveillance technologies, including, but not limited to, the following:
  —closed circuit television (29 agencies);
  —night vision systems (22);
  —miniature transmitters (21);
  —electronic beepers and sensors (15);
  —telephone taps, recorders, and pen registers (14);
  —computer usage monitoring (6);
  —electronic mail monitoring or interception (6);
  —cellular radio interception (5);
  —pattern recognition systems (4); and
  —satellite interception (4).
—About 25 percent of Federal agency components responding (36 out of 142) report use of computerized record systems for law enforcement, investigative, or intelligence purposes.[5]

Sample agencies in this last category included the FBI, Treasury, and the Immigration and Naturalization Service—as might be expected.

A second report, *Management, Security, and Congressional Oversight*,[6] was less concerned with privacy and other civil liberties issues, focusing instead on five major areas: (1) management of the new technologies, (2) system security and computer crime, (3) decision support, (4) information dissemination, and (5) possible Congressional oversight. This clearly represents the management side of the issue(s).

The third report in the series, *Electronic Record Systems and Individual Privacy*, returns to the civil liberties aspects but also focuses on possible policy changes, including possible amendments to laws such as the Privacy Act of 1974. In this, OTA identifies a range of policy options open to Congress beginning with a do-nothing approach (clearly not recommended):

1. Congress could do nothing at this time, monitor Federal use of information technology, and leave policymaking to case law and administrative discretion. This would lead to continued uncertainty regarding individual rights and remedies, as well as agency responsibilities.

Additionally, lack of congressional action will, in effect, represent an endorsement of the creation of a *de facto* national database and an endorsement of the use of the social security number as a *de facto* national identifier.[7]

The second option discussed is an ambitious one. I quote it in full because it covers so much ground:

> 2. Congress could consider a number of problem-specific actions. For example:
> —establish control over Federal agency use of computer matching, front end verification, and computer profiling, including agency decisions to use these applications, the process for use and verification of personal information, and the rights of individuals;
> —implement more controls and protections for sensitive categories of personal information, such as medical and insurance;
> —establish controls to protect the privacy, confidentiality, and security of personal information within the micro-computer environment of the Federal Government, and provide for appropriate enforcement mechanisms;
> —review agency compliance with existing policy on the quality of data/records containing personal information, and, if necessary, legislate more specific guidelines and controls for accuracy and completeness;
> —review issues concerning use of the social security number as a *de facto* national identifier and, if necessary, restrict its use or legislate a new universal identification number; or
> —review policy with regard to access to the Internal Revenue Service's information by Federal and State agencies, and policy with regard to the Internal Revenue Service's access to databases maintained by Federal and State agencies, as well as the private sector. If necessary, legislate a policy that more clearly delineates the circumstances under which such accesses are permitted.[8]

The third option ends with an old recommendation of an oversight board or commission:

> 3. Congress could initiate a number of institutional adjustments, e.g., strengthen the oversight role of OMB, increase the Privacy Act staff in agencies, or improve congressional organization and procedures for consideration of information privacy issues. These institutional adjustments could be made individually or in concert. Additionally or separately, Congress could initiate a major institutional change, such as establishing a Data Protection or Privacy Board or Commission.[9]

The final option is simply a proposal for more study—though in a broader context:

> 4. Congress could provide for systematic study of the broader social, economic, and political context of information policy, of which information privacy is a part.[10]

Obviously these three reports do not address all the issues that might be addressed under the heading of new electronic technologies and civil liberties. For example, they do not consider problems associated with these same technologies in the hands of state government agencies, of corporate managers, of owners of small businesses, of criminal enterprises, or of foreign agents. Nor, even in the limited area of federal agencies, do they address all the issues that concerned citizens want to see addressed. But at least the first and third reports represent a good start—and a satisfactory narrowing of the issues for discussion here.

I end this introduction to the problems with still another quotation. It is one of the recommendations in the first (1985) OTA study—namely, that "Congress could bring new electronic technologies and services clearly within the purview of Title III of the Omnibus Crime Control and Safe Streets Act." This could be done in several ways, the authors say, by:

—treating all telephone calls similarly with respect to the extent of protection against unauthorized interception, whether analog or digital, cellular or cordless, radio or wire;
—legislating statutory protections against unauthorized interception of data communication;
—legislating a level of protection across all stages of the electronic mail process so that electronic mail is afforded the same degree of protection as is presently provided for conventional first class mail;
—subjecting electronic visual surveillance to a standard of protection similar to or even higher than that which currently exists under Title III for bugging and wiretapping.[11]

A question can be raised about the likelihood of these reforms, but in 1986 Congress did, as recommended in this report, pass an Electronic Communications Privacy Act.[12] Nonetheless, the need for reform seems still to be great.

COMPUTER ETHICS AND THE LIKELIHOOD OF REFORM

At the beginning of this chapter, I quoted philosopher Deborah Johnson to the effect that ethics in the area of computer use may require "responsible behavior on the part of . . . those who understand and work with computers." Johnson's book, *Computer Ethics*, takes up

> issues of responsibility and liability surrounding computers. Here the emphasis is on duties of computer professionals in marketing computer products, and their liability for mistakes and malfunctions in computer programs. The ethical problem is to assign duties and liabilities in a way that is fair both to computer professionals and to consumers of computer

products.[13]

Other issues explored in Johnson's book include effects of the increased use of computers—for example, issues of privacy or alleged centralization or decentralization of power, including computer professionals' responsibilities to control the effects of computerization. Johnson thinks they have a limited responsibility to help protect privacy, to become knowledgeable about the social impacts of computers, and to bring these impacts—especially if they are perceived to be bad—to the attention of management. The final chapter in the book is devoted to what Johnson says is a very complex question: who owns computer programs and how the ownership of programs should be regulated. Here she wades through legal thickets of patent protection, copyright, and trade secrets and concludes that "the status quo is acceptable" or at least "is not morally unacceptable."

Johnson's *Computer Ethics* is something of a pioneering effort and she begs critics to be kind, to "use this book as a starting place, rather than as the final word."[14] One reviewer, Carl Mitcham, has failed to heed this plea for mercy. Because Johnson concentrates almost exclusively on what computer professionals should or should not do based on contemporary deontological or utilitarian theories in applied ethics, Mitcham accuses her of too narrow a focus.[15] Mitcham would have Johnson delve into deeper issues, and he chides her for saying that while "whether or not computers really duplicate human thinking is an enormously interesting and complex matter," it "is not ethical in character." Mitcham prefers Sherry Turkle's social-psychological approach in *The Second Self: Computers and the Human Spirit.*[16] There Turkle's primary focus is "the subjective computer . . . the computer as it affects the way that we think, especially the way we think about ourselves."

In this chapter, I side with Johnson rather than Mitcham. I am more interested here in questions of ethics and social responsibility than in epistemological or metaphysical issues about artificial intelligence. But I do worry that Johnson's approach to computer ethics does not go nearly far enough in dealing with the most pressing ethical and social problems of our computer and electronic age. What I am concerned about is not so much whether computer professionals recognize their ethical and social responsibilities, but whether they act in accord with those responsibilities and do something about the social problems.[17]

COMPUTER SCIENTISTS' ACTIVISM

The group that I want to concentrate on here is Computer Professionals for Social Responsibility. What these people maintain is that:

> We should expect that the intimate details of our private lives enjoy the same protection whether big business or big government is the custodian. Absent clear privacy safeguards, we are left at the mercy of a rapidly evolving technology and an industry that can say little more than, "trust us."[18]

Most people concerned about these issues are probably familiar with CPSR, but it is worthwhile to say something about its history and activities.

Social responsibility had been a concern among computer scientists earlier—recall, for example, Joseph Weizenbaum's *Computer Power and Human Reason*[19]—but CPSR came into being as a formal organization only in 1982. There are now over twenty chapters, and membership is at three thousand or so.

According to CPSR's promotional brochure, it has been active in testifying before committees of the U.S. Congress, in contacting the media and alerting the public about electronic invasions of privacy and other infringements of civil liberties, in promoting forums for the public discussion of issues such as the software requirements of the Strategic Defense Initiative (Star Wars) as well as privacy issues, in watchdogging the FBI's efforts to expand crime information records, and in publishing civil-liberties-related materials. One example of their publications is a *Special Report on Computers and Elections* (published jointly with the Urban Policy Research Institute), intended for use during the 1988 U.S. Presidential election. In several of these efforts, CPSR has formed coalitions with other groups such as the American Civil Liberties Union.

As an example of CPSR activity, Mark Rotenberg of the Washington, D.C., office, made a presentation at a "New Technologies and Civil Liberties" conference run by the American Civil Liberties Union Delaware chapter in May, 1990. There he both reported on CPSR's advocacy efforts before a U.S. House subcommittee hearing on the Data Protection Act of 1990 and joined with Janlori Goldman of ACLU in making a strong case against corporate misuses of information on individual consumers—for example, selling illegally gotten buying profiles for telemarketing purposes.

As Rotenberg makes his case, it is obvious that CPSR is interested in doing more than just making polite appeals to academic audiences. The group wants to generate a public, activist constituency to fight against misuses of electronic technologies and invasions of people's privacy. Here is an example of technically trained professionals attempting to go beyond their own narrow interests—even beyond a narrowly-defined requirement of social responsibility on the part of computer professionals—to make common cause with other activists to *get something done*, and not just talk about threats to civil liberties.

A better-known example of this sort of activism by CPSR is the refusal of some of its members—along with a great many other scientists and technically trained individuals—to accept research funds for work on the Strategic Defense Initiative.

The pledge not to accept funding for SDI research seems to have begun with physicists, notably at Cornell University and the University of Illinois in 1985.[20] More than one version of the pledge was circulated, but here is the most popular version:

Anti-SDI Pledge

> We, the undersigned scientists and engineers, believe that the Strategic Defense Initiative (SDI) program (commonly known as Star Wars) is ill-conceived and dangerous. Anti-ballistic-missile defense of sufficient reliability to defend the population of the United States against a Soviet attack is not technically feasible. A system of more limited capability will only serve to escalate the nuclear arms race by encouraging the development of both additional offensive overkill and an all-out competition in anti-ballistic-missile weapons. The program will jeopardize existing arms control agreements and make arms control negotiation even more difficult than it is at present. The program is a step toward the type of weapons and strategy likely to trigger a nuclear holocaust. For these reasons, we believe that the SDI program represents, not an advance toward genuine security, but rather a major step backwards.
>
> Accordingly, as working scientists and engineers, we pledge neither to solicit nor accept SDI funds, and we encourage others to join us in this refusal. We hope together to persuade the public and Congress not to support this deeply misguided and dangerous program.[21]

By early 1986, over four thousand scientists, engineers, and students had signed the pledge (in one form or another).

At about the same time, CPSR began circulating a similar petition calling SDI "technically infeasible with known or anticipated techniques" any time before the year 2000. One report indicates that it was signed rather quickly by dozens of computer experts, including chairs of prestigious university computer science departments.[22] Later on, CPSR soft-pedaled its pledge drive but continued to hammer away at aspects of the SDI initiative that seemed to threaten escalation of the nuclear arms race. (I treat nuclear protests in chapter 9.)

It seems finally not to have been these refusals on the part of physics and computer science researchers to accept funding that slowed the drive to promote SDI. That came about as a result of developments in Eastern Europe and the Soviet Union, as much as anything. Nonetheless, the CPSR and physics dissenters certainly contributed a great deal to public debate of SDI, and their activism in general shows more promise of getting something done about problems of our computer age than calls for more ethical behavior on the part of computer professionals—whether those calls emanate from applied ethicists or from authorities in computer professional societies or managers of computer firms or government agencies. I would not disparage calls for better ethics codes, better enforcement, or an "ethical climate" in computer and electronic worksites. But if the problems of our electronic age are as serious as the Office of Technology Assessment studies suggest—and the OTA conclusions are mild when compared with the warnings of more radical critics[23]—then it seems that action is what is called for.

9

Nuclear Experts and Activism

The scientific and technological infrastructure created for the Manhattan Project underwent "civilianization" after the end of World War II. A substantial piece of the infrastructure was directed under the 1946 Atomic Energy Act to promote the "peaceful use of the atom."

—John Byrne, Steven Hoffman, and Cecilia Martinez

There is indeed a connection between production of nuclear weapons and generation of electricity by means of atomic energy. Not only are there unalterable historical and technical links between the two applications of the technology, but also commercial nuclear power is a necessary condition (and close to being a sufficient one) for military applications of fission technology.

—Kristin Shrader-Frechette

Some people would object to my linking nuclear weapons with the nuclear generation of electricity in this chapter. Defenders of nuclear power—for example, Samuel McCracken and Bernard Cohen[1]—chide opponents for linking the two, even accusing them of irrational fears that nuclear plants can blow up like nuclear bombs. I lay no claim to expertise on what can or cannot happen to a nuclear power plant, but in any case my focus is not on the possibility of a blow-up (I assume the likelihood is close to zero) or even of a meltdown (a possibility with some likelihood). My focus is on the close connection, historically, between the communities of scientists and engineers who have done the research, designed and run both sorts of facilities, and been the staunchest defenders of the safety of nuclear power.

John Byrne and colleagues, citing the standard history of the early days of atomic research in the United States by Richard Hewlett and Oscar Anderson,[2] quote these figures:

> To accomplish the task [of the Manhattan Project], a giant industrial complex was erected consisting of thirty-seven installations in nineteen states and Canada, 254 military officers, approximately 4,000 scientific and technical personnel, and a $2.2 billion [budget].[3]

It is to this organizational structure that they refer when they say that a "substantial piece of the infrastructure" was redirected to the "peaceful use of the atom" after 1946.

A very high concentration of these people, associated with the DuPont Company, live and work in Delaware—and have, before, lived and worked at Savannah River, on the South Carolina-Georgia border, or at Hanford, Washington. My personal experience is limited to conversations with dozens of these people.

ARGUMENTS FOR LINKING NUCLEAR TECHNOLOGIES

Kristin Shrader-Frechette, an outspoken philosopher-critic of nuclear technologies, makes one of the strongest cases for linking nuclear weapons with the nuclear generation of electricity. Shrader-Frechette formulates her argument under three headings: historical, technical, and philosophical links between the two technologies.[4]

The argument for a historical link is easiest to make. Shrader-Frechette cites Carl Walske, a physicist and president of the Atomic Industrial Forum, as saying on many occasions that military expenditures were ultimately responsible for the development of commercial atomic energy. She claims that, between the late 1940s and the mid-1960s, the U.S. government spent $100 billion to develop the first nuclear reactors—fifty times as much as was spent on producing the atomic bombs dropped on Hiroshima and Nagasaki. Why?

> The "Atoms for Peace" program was initiated and continued, in large part, because it provided a nonwarlike rationale for continuing the development of nuclear fission. Prodded on both by the escalating Cold War and by hope in the "peaceful atom," the U.S. government pushed development of commercial reactors for generating electricity so that it could obtain the weapons-grade plutonium as a by-product.[5]

Although this view is occasionally challenged, I have not encountered a single scientist or engineer—among the dozens and dozens I have known who worked either at Savannah River or Hanford—who would deny the link. In fact, the link seems obvious, natural, and defensible to almost all of them precisely because they are so much in favor of *both* kinds of applications of nuclear technology. Even the small percentage whose anti-war activism later led them to oppose nuclear weapons often justify their earlier work *because* it had non-military spinoffs in the nuclear power industry.

Shrader-Frechette goes on to establish a technical link between military and commercial nuclear technologies: "Enrichment plants existed for the bomb effort, and their continued operation could be justified if they were also used to make fuel for reactors. For this reason, the U.S. nuclear technology has been built around a water-cooled, enriched-uranium design."[6] Here she may exaggerate somewhat; surely there were other reasons for the choice, not least of which was the experience gained by reactor-design scientists and engineers, for example, in companies like General Electric and Westinghouse.

Against Shrader-Frechette's intentions, it can be claimed that the *only* link between weapons reactors and commercial reactors is technical. Joseph Morone and Edward Woodhouse, in fact, make precisely this claim: "In sum, technical factors almost completely determined the early reactor choices. By far the dominant consideration was the perceived need for breeding, arising out of the apparent scarcity of uranium."[7] However, Morone and Woodhouse are not focusing on the same issue as Shrader-Frechette—linking military and commercial uses of nuclear technology—but on the relative neglect, by early designers, of issues concerning the safety of the public.

In any case, Shrader-Frechette goes on to support her technical-link conclusion with four arguments: the first is based on laser-separation technology; the second on the dual capability of the reactors, to produce either electricity or weapons-grade plutonium; the third on the commonality of skills needed for the two kinds of program; the fourth and last on the organizations involved.

It is Shrader-Frechette's last two arguments that I would emphasize. All that government money funded an enormous complement of nuclear physicists, engineers, and technicians trained roughly between 1950 and 1970. These people could, and some of them did, move easily back and forth between military and civilian projects, utilizing the same technical and organizational skills in both.

Shrader-Frechette's third set of arguments are philosophical, and they are the most complex—as well as the most controversial. She summarizes them under two terms, controllability and security, and she argues that advocates regularly commit the same fallacies in defending both programs. They claim that both military and commercial nuclear technologies are controllable; she says they are not. Advocates similarly claim that both technologies contribute to national security—something that Shrader-Frechette argues they do not. For my purposes here, these pro and con arguments are less interesting than the historical links and the links based on skills and organizational ties.

When I first got interested in the nuclear power issue, I had a reasonably open mind about it. My main opposition was to the nuclear arms race. At first, I could find few of my scientist or engineer friends who would even venture to be on a panel discussing the pros and cons of nuclear power; however, gradually more and more of them came at least to the point of entertaining doubts about nuclear power. And that is the topic I address next.

NUCLEAR POWER AND ITS OPPONENTS

Shrader-Frechette, though she claims to be objective and even-handed in her assessment, has acquired a reputation as an opponent of nuclear power.[8] Dorothy Nelkin, on the other hand, has managed to convince more of her readers that she is fair to both sides in her chronicling of opposition to nuclear power both in the United States and in Europe.[9] Nelkin also emphasizes certain features of the opposition that are crucial to my argument in this book.

Nelkin's first major contribution to the debate over nuclear power was her book, *Nuclear Power and Its Critics*.[10] Focusing centrally on opposition to a plant on Cayuga Lake proposed by the New York State Electric and Gas Corporation in 1967, Nelkin emphasizes a complex set of problems and questions:

> The case brings into focus relationships among scientists, public agencies, citizens' groups, and policy makers. . . . A major issue in the dispute was the character of scientific evidence itself and its interpretations for policy purposes. What are the obligations of scientists with respect to interpretation of their data? . . . How does the behavior of opposition groups affect decisions concerning environmental issues?[11]

One important factor Nelkin focuses on is the conflict among technical experts this controversy elicited:

> The . . . reaction to the Bell Station plans must be considered in relation to the location of the site, 12 miles from Cornell University. Here was a high-powered cluster of scientists and engineers with concerns for the preservation of a major resource in their community.[12]

But by no means all of the Cornell scientists and engineers involved in the controversy were on the side of the opponents. To summarize, they represented a very broad spectrum, from supporters (typically more cautious than the company's technical teams) to people who ended up joining the citizens movement to stop the building of the plant.

This is the next important factor I would like to emphasize, the collaboration of scientists and engineers with the opposition groups. In the early days of the controversy, Nelkin reports, the technical people involved were not completely opposed to building the plant, and focused mainly on preserving the ecology of Cayuga Lake. But in the later stages, when national movements in opposition to nuclear power had gained headway, a number of the scientists and engineers joined the more vocal critics, emphasizing not only ecological problems but risks associated with nuclear plants and nuclear waste disposal problems. In short, as with my scientist and engineer friends, many started out with a posture of technical neutrality but ended up as activists in opposition not only to this one plant but to nuclear power generally. (In their view, scientists and engineers for the electric company had, long before, abandoned their claim to scientific objectivity and technical neutrality.)

The Bell Station/Cayuga Lake controversy came nearly a decade before the near-catastrophe at Three Mile Island and the catastrophe at Chernobyl. Meantime, opposition to nuclear plants had already become an international phenomenon. In this changed environment, it is no longer at all unusual to find scientists and engineers, even those trained in nuclear fields, who are critical of nuclear power, whether in particular cases or in general. Neither is it at all unusual for critical scientists and engineers to join with other anti-nuclear activists.

CAN OPPONENTS OF NUCLEAR POWER SUCCEED?

John Byrne and colleagues, borrowing "autonomous technology" ideas from Jacques Ellul, make a strong case for the claim that Three Mile Island and Chernobyl represent no more than temporary setbacks—that, after some troubling times, the nuclear power project will regain its lost momentum. They cite Llewellyn King, editor of *Energy Daily*, as saying,

> The nuclear industry's place is secure; its present is difficult. By the mid-1990s alternative energy systems, such as solar, which have been found wanting, will not be suitable or adequate and we are going to go nuclear. The whole world is going to go nuclear.[13]

They confirm this with a quote from Rémy Carle, construction director for Electricité de France, to the effect that abandoning the nuclear power project "would involve such great tensions on the energy market that [it] would endanger world peace."[14]

These might be seen as pious hopes of diehard defenders of nuclear power, and the case that Byrne et al. make does not ultimately rest on such predictions. Their case rests on what they call the "nuclear consortium"; the Atomic Energy Commission's "method for achieving its goal of reactor commercialization," they say, was the same as that of the Manhattan Project. That is, it centered around a combination of "Big Science, Big Industry and Big Government."

> The AEC was the home of Big Science, its laboratories serving as the "factories" for technology development. . . . The agency relied on a close circle of corporate and university giants to carry out the mission. Through the 1960s, the AEC distributed over half of its expenditures to just five companies—DuPont, General Electric, Union Carbide, Bendix and Sandia, and two schools—the University of Chicago and the University of California.[15]

These organizations could be trusted to get the job done efficiently with the "tight control needed in top secret organizations."

But a consortium of these experts was not all; they had an aim of a "suitably centralized technologic result." Ultimately, they expected to dominate the power industry with gigantic nuclear technology. The Ellulian inference Byrne et al. draw from this is that,

> The electrical network as it developed in the 20th century was a force for technocratic rule, not democratic governance. The Nuclear Project . . . adding political secrecy and security to the institutional apparatus . . . sought and obtained institutional autonomy and accumulated power at the expense of democratic social institutions.[16]

And it is this system that they quote experts as saying is now being successfully exported all over the world, even after Three Mile Island and Chernobyl.

Given the power of this government-industry consortium—along with its penchant for secrecy, its goal of universalization, and the prediction that abandoning nuclear power would be a recipe for political instability or even war—what are the *real* prospects for those anti-nuclear activists who see nuclear power as both environmentally risky and undemocratic? Byrne et al. would suggest that their chances of doing anything more than slow the nuclear juggernaut are slim at best.

I do not see the picture as being as bleak as that. It seems to me that Byrne and his colleagues—who are or have been my colleagues at the University of Delaware—are themselves evidence that their cause is not hopeless. Whether or not the nuclear project is ultimately brought to a halt may not, indeed, be the issue. What seems to me the larger issue is whether or not nuclear decisions can be made democratically, and here the prospects are at least a little bit brighter.

Morone and Woodhouse, in *The Demise of Nuclear Energy?: Lessons for Democratic Control of Technology*, make precisely this point. Their basic assumption is that the form a technological enterprise takes does not have to be incompatible with public preferences. About the nuclear enterprise in the United States, they say this:

> There was nothing preordained about the exact type of technology and regulatory system that we inherited. True, the head start that light water reactors enjoyed made their dominance likely. But they did not have to be large reactors, nor did they have to rely primarily on prevention for the last line of defense. Reactors did not have to be so close to cities, nor sold in such large numbers so quickly.[17]

Morone and Woodhouse pinpoint the exact time in which "the system became locked in . . . in just three years in the mid-1960s."[18] Though this might have been predicted, given appropriate knowledge about the key decision makers and about such social conditions as the Cold War and widespread public support at the time for technological development, Morone and Woodhouse think a go-slow approach might still have been possible at the time.

However, what Morone and Woodhouse are really interested in showing is how a *new* democratic approach is possible today. Their final chapter is devoted to showing just this, and they conclude: "Just as nuclear power technology was initially shaped by political and economic forces a generation ago, so it can be reshaped [into an acceptable form] if a political majority becomes willing to undertake the task."[19]

The weakness in Morone and Woodhouse's analysis seems clear to me in their conclusion. They ask: "Now that they have won the battle, will those who opposed large-scale nuclear power turn away from energy issues altogether?"[20] This does not seem to give anti-nuclear activists much credit; it seems to assume that, while nuclear advocates might scale down the size of plants and make them safer, and while the public might be persuaded to accept a more limited form of nuclear power, the activists will remain stuck in their oppositional stance of the 1970s and 1980s.

Presumably Morone and Woodhouse wrote their book, at least in part, to persuade anti-nuclear activists to give up their all-or-nothing oppositional stance—especially in light of the potential environmental catastrophes associated with current alternatives to nuclear power. It seems to me that anti-nuclear activists can be just as adaptable as anyone else in this debate; that many of them (at least those I know well) are every bit as attuned to broader environmental issues as Morone and Woodhouse; and, most important of all, that new activists from the nuclear science and engineering communities can make a great difference as the public debate over nuclear power moves into the 1990s, and on into the twenty-first century. Diehard advocates of nuclear power (and sometimes Morone and Woodhouse seem to support that group) and diehard opponents of any and all of its forms do not exhaust the possibilities of activism on this issue. Activism can be flexible and intelligent—and, most important, it can be democratic in the best sense.

ETHICS AND NUCLEAR WEAPONS RESEARCH

I want now to turn to another facet of the nuclear issue. Activism has been as much a part of public discussion of nuclear weapons as of public discussion of nuclear power, but I focus here on the much neglected issue of ethics and military research, including research on nuclear weapons.

Along with scientist Philip Siekevitz, philosopher Carl Mitcham was instrumental in getting the New York Academy of Sciences to hold what was billed as the first major conference on that topic in January 1989. In Mitcham's introductory essay for the proceedings volume emanating from the conference, he outlines three ways ethics can be related to military research: in terms of actions, such as whistleblowing or quitting or refusing to do research on weapons; in terms of deciding what is morally right or wrong in such cases; and in terms of rationally justifying such conclusions by way of philosophical/ethical systems.[21] As it turned out, the conference tended to emphasize the first and second sorts of concerns much more than the third concern of abstract philosophizing.

One thing that is missing in the Mitcham and Siekevitz volume is a sense of the history of scientific involvement in military research. Daniel Kevles, in his well known study of one scientific community, *The Physicists*, demonstrates that only a small percentage of physicists objected to doing research for the military in either World War I or World War II—and even in those cases the reason was often not an objection based on moral principles but a matter of science policy.

With respect to World War I, Kevles notes:

> Some scientists grumbled at the [new National Research] Council's ambitions, complaining that it was trying to impose a dangerous centralization on American research. Others agreed with James McKeen Cattell, a pacifist, who indicted the Council as "militaristic."[22]

On the same note, Kevles later writes:

> In 1938, almost 1,300 American scholars and scientists, representing 167 academic institutions and ranging ideologically from Robert A. Millikan to Dirk J. Struik, the mathematician who also edited the Marxist *Science and Society*, issued a manifesto. The document condemned the fascist suppression of science, castigated Nazi racial theories, asserted the legitimacy of modern theoretical physics, and ringingly concluded: "Any attack upon freedom of thought in one sphere, even as nonpolitical a sphere as theoretical physics, is an attack on democracy itself."[23]

Again Kevles acknowledges a small number of dissenters:

> Various scientists, largely on the political left, some of them pro-Communists, others pacifists, disapproved of the preservation of intellectual freedom by military force.

It is worth noting what the dissenters proposed in place of doing military research:

> [The dissenters] proposed instead to ease the material wants of mankind, keep the international scientific community an apolitical beacon of peaceful cooperation, and, in the last resort, "to go on strike," as *Scientific American* urged, if the government demanded the use of their technical talents for military purposes.[24]

In the end, the number of dissidents was so small that, as Kevles notes, the Second World War could be called "a war of physicists"—as "World War I had been a chemists' war."[25]

In the Mitcham-Siekevitz volume, Douglas MacLean offers a moral rationale for this attitude on the part of scientists and engineers: "We should see the bomb-maker," he says, "eagerly and enthusiastically dreaming up ever more deadly weapons, in [the] context [of the entire defense establishment]." Instead of a negative view, MacLean asserts: "We should . . . see this person [or the engineer in a national laboratory confidently optimistic about the Strategic Defense Initiative] as bringing enthusiasm and dedication to his work, traits that we would normally praise." And MacLean concludes: "I suspect it is often wrong to demand too great a moral sensibility on the part of each of these individuals in examining each project they undertake." MacLean reserves his moral strictures for "the entire defense establishment—the military-industrial-research complex," bringing "the executive, the politician, the banker, and investor" into the picture.[26]

While thus exempting most researchers on weapons systems from moral blame, MacLean does admit this much: "The engineer who blows the whistle on some lucrative but foolish enterprise is a noble person, and the ones who let loyalty override good sense are condemned and occasionally punished."[27]

Rosemary Chalk, on the other hand, stresses the role and importance of what she calls scientific "conscientious objectors." Chalk's first example is an American Quaker pacifist, A. J. Muste, who urged on Albert Einstein and his Emergency Committee of Atomic Scientists—who were urging public support for scientists' efforts to establish international control of nuclear weapons—that scientists make an "expression of moral commitment," that they "renounce publicly any further involvement in building these weapons."[28]

Chalk admits that "the idea of drawing lines in some absolute fashion on military research . . . has received little enthusiasm or collective support within the scientific and technical community."[29] However, she notes how some leading scientists and engineers do support a qualified resistance, "proposing that scientists should not participate in the development of particularly destructive weapons of war, most commonly with reference to nuclear, biological, or chemical weapons."[30] Three outstanding examples Chalk cites are Peter Kapitza, a Soviet physicist, the mathematician founder of the field of cybernetics, Norbert Wiener, and Victor Paschkis, an engineer largely responsible for organizing both the Society for Social Responsibility in Science (SSRS) and the Society for Social Responsibility in Engineering (SSRE).

Chalk ends this section of her paper on a cautionary note, reminding us that "the idea of conscientious objection in science" has not won broad support. She even quotes a leading scientist as saying, "I thought they [objectors to the development of the hydrogen bomb] were quite off their rocker, frankly."[31]

An outstanding example of a dissenter who matches Chalk's model is Stephen Unger, a computer scientist who has been very active in moving the Institute of Electrical and Electronics Engineers (IEEE) in the direction of supporting whistleblowers. In his contribution to the Mitcham and Siekevitz volume, Unger limits himself to the question whether engineering ethics has any relevance to the issue of whether or not engineers can, with a good conscience, work on weapons projects. Unger notes that even among engineers with qualms about doing so, their motivations—and the value systems their motives are based on—are remarkably diverse. The only thing all might agree upon is this: "whether the project is likely to lead, on balance, to a net loss or gain in terms of human life, health, and welfare."[32] And even then, as Unger says he learned by sad experience, there is no guarantee that a conscientious refusal—or a switch to some other project thought to be less harmful—will not backfire, with one's work being used, despite the protest, for some military purpose. Unger concludes:

> Although it is sometimes difficult to determine whether or not a project is, on balance, harmful or beneficial, I do not agree . . . that the hard part of being moral is in figuring out what is moral. There is ample evidence that, for engineers, doing what is obviously right can sometimes be extremely costly.[33]

What this suggests is the need for more than reliance on engineers' being ethical. For me (and I think for Unger), what is needed is organizational change—which almost always requires public pressure. I believe, as I will be saying shortly, that we need activism that will generate public pressure for ethical changes in weapons research. But at the moment there seems to be no great swell of activism on this issue.

Where there has been a swell of activism—indeed, a series of activist swells—is in opposition to nuclear weapons (rather than to doing research on them). An intriguing study of the various antinuclear activist movements, but especially of the Nuclear Freeze movement of the 1980s, has been done by Frances McCrea and Gerald Markle. Though it constitutes an aside, I want to devote a few paragraphs to McCrea and Markle's study.

Undoubtedly because of their radical anti-capitalist stance, McCrea and Markle end their book on an avowedly pessimistic note. The Freeze, they say, almost inevitably fell into the hands of people who would adopt strategies that were doomed to fail—or at least to fall far short of the stated objectives of the movement, namely of eventually eliminating nuclear weapons. Identifying three factions in the Freeze movement, McCrea and Markle say this about the ultimately dominant faction:

> The strategy of peace liberals was to exert as much political pressure as

> possible, principally through electoral politics. . . . In order to attract mass support, social movement leaders must employ tactics, such as education and participation in electoral politics, that are acceptable to a middle class constituency. . . . Such support [however] is likely to be withdrawn during hard times—the times that such support is most needed.[34]

McCrea and Markle clearly favor the remnants of the New Left in the Freeze movement, along with a second faction of radical (especially religious) pacifists, yet they are realistic enough to recognize that "such organizations are, and always have been, on the fringes of American politics; and their ability to influence public policy has always been minimal."[35]

Despite their pessimism, McCrea and Markle admit that the Freeze movement did accomplish a great deal:

> Although Freeze victories are largely symbolic and in substantive terms have not altered the arms race, the Freeze Movement has achieved important goals. Technocratic consciousness has been challenged. Nuclear weapons issues are no longer seen as the exclusive domain of military and political experts, but have become a topic of public discussion. Most importantly, the movement has defined a new social problem and has deepened the legitimation crisis faced by the ruling elites.[36]

Interesting, in McCrea and Markle's analysis of the Freeze movement, is the ambivalent role of scientists. On one hand, they say: "Of all modern social movements, scientists have played a major role in only one—the protest against nuclear weapons."[37] However, "Nuclear scientists, even in protest, have emphasized instrumental reasoning and technical discourse . . . [and have] consistently eschewed 'emotional' appeals such as . . . *New York Times* ads . . . as simplistic."[38] Nonetheless, McCrea and Markle believe that scientists and other technical people can play a more activist role, and they predict that they will make a difference—for example, in the environmental movement (which I take up in the next chapter)—if they can learn to be more politically savvy.[39]

Returning to the issue of the ethics of nuclear weapons research, my own contribution to the Mitcham and Siekevitz volume emphasized the need for scientific activism.[40] I point out there that in my advocacy of public-interest activism by engineers and scientists—aimed at making their institutions more socially responsible—I have not elsewhere faced the challenge of addressing my exhortations to people who might be faced, at every turn, with the challenge that they might be undermining national security. (At the extreme, of course, they have been accused of more—of treason, or at least of "playing into the hands of our enemies.")

One of the main tenets of my approach to ethics, as should be obvious by now, is an advocacy of the tolerance and openness associated with a respect for civil liberties. But this too confronts a long tradition, even in the best of administrations, of suspending civil liberties in the name of national security. (Presumably such suspensions are justified as extensions of similar suspensions made in times of war.) If my readings of the defenses of such suspensions of civil liberties are correct, they imply that the Constitution's delegation of the powers of commander in chief to the president supersedes the Bill of Rights. These defenses include assertions that the Constitution allows, at the least, temporary suspensions of some provisions in the Bill of Rights "for the common defense." It might, by further extension, then be said that research in the name of national security enjoys a similar set of privileges. Acceptance of this extension, however, would compromise the right of dissent in research communities working on national security projects.

Perhaps, someone might say, dissenting views could be circulated outside the military. For this to happen, citizens in public debate (perhaps some of them military researchers) would have to raise issues of social responsibility with respect to particular weapons development projects. Ordinary citizens, however, seldom know much about weapons development projects until they are nearly completed. Furthermore, the same problem affects public dissent with respect to military developments as impedes dissent within the ranks of military researchers. Public discussions of weapons policies are almost always couched in the language of threats to national security—or even survival. (This language is employed by those on the left as well as by those on the right; that is, doves also worry about the survival of the nation, though they are usually thinking of the survival of a national commitment to the Bill of Rights and other civil liberties.) Is it possible to maintain an atmosphere of civil discussion when each side believes that losing the argument would threaten the defense of a free society or the survival of the free world in the face of a nuclear catastrophe?

Although some may think it impractical to call for civility in discussions of society-threatening issues, I believe that civilized public debate of these issues is essential. The Mitcham and Siekevitz volume discusses the moral pros and cons of research for the military, and such discussions are long overdue. Also, this volume includes contributions by military researchers, and such contributions are particularly welcome in debates on these issues.

I also believe, however, that sometimes ethics demands more than polite discussion. If particular groups, and there is no reason that military researchers cannot be among them, feel deeply that specific weapons development projects are unethical or are opposed to the values of a free society, and if normal discussions seem to be getting nowhere in changing the course of development, it may well be necessary to mount public protests, including peaceful and nonviolent demonstrations. If the protestors are particularly incensed and the weapons development policy seems particularly outrageous, I can even imagine another mass political movement of the sort that opposed the war in Vietnam.

In such cases, where people feel that ethics demands public confrontation of military policy, the protestors, if they have any sense, should recognize that defenders of the particular policy will inevitably also come forward. The protestors, as happened during the Vietnam War, may even be accused of treason. And the defenders of military policy are also likely to dress their arguments in the clothing of flag and country—even, on occasion, defending their views with strong ethical arguments.

Polite discussion may be preferable to public protests and counterprotests, but in my view ethics sometimes demands public protest, even national soul-searching, if particular policies seem egregiously opposed to national values.

This chapter has tied together two issues: the seemingly successful movement to halt nuclear power in the United States—a movement that, until recently, did not include many scientists and engineers, especially those in nuclear specialties—and the nuclear weapons issue. Quite a few scientists have gotten involved, over four decades, in movements to control nuclear weapons. What I want to emphasize here is the need for a broader social responsibility movement among people doing research for or under the auspices of the military. In the end, I believe they can be effective, but only if they join with other activists—including those in opposition to nuclear power—along with activists in the environmental movement, to which I turn next.

10

Environmentalists and Environmental Ethics

[There are] occasional but persistent complaints by some readers [of Environmental Ethics] that feature articles are too obscure and difficult to read. . . .

It is unrealistic, I believe, for anyone to expect to understand articles on environmental ethics meeting minimum professional standards without knowing something about philosophy as an academic subject.

—*Eugene C. Hargrove*

I wish to respond to points raised about the philosophical value of papers [in Environmental Ethics] and their relevancy to environmental affairs in [editor Hargrove's] recent editorial. . . . [A] suggestion to improve both philosophical value and relevancy . . . to environmental affairs would be for philosophers and environmental professionals to work more closely with one another in mutual projects.

—*John Lemons*

This exchange between Eugene Hargrove[1] and John Lemons[2] appeared just a few years after the academic journal, *Environmental Ethics*, began publication in 1979. While Hargrove initiated the exchange with his defense of "professional standards" for environmental ethicists, Lemons countered with his claim. Academic rigor is not the problem; it is lack of relevancy.

Environmental ethics as a focus of attention within applied philosophy is little more than a decade old. However, in that short time, enough has been published for us to reflect on the tenor of the literature in the field. By now there are at least a half dozen books that attempt to provide a theoretical grounding for the field. And there are several anthologies that collect a wide variety of contributions to debates on foundations, including whether or not the field requires a radical restructuring of traditional approaches to ethics—even of approaches to the rather new tradition of applied ethics.[3]

The academic rigor versus environmental relevance discussion, above, reminds me of John Dewey, who argues forcefully—for instance, in *Reconstruction in Philosophy* and *Liberalism and Social Action*[4]—that philosophy's focus ought to be not on academic but on real-life social problems.

Reduced to its simplest terms, Dewey's argument runs something like this:

(1) Since the beginnings of Western philosophy, those philosophers who have merited appellation as great have all made significant contributions to the solution of the moral and social (and occasionally also the political) problems and conflicts of their historical epochs. (Indeed, historical epochs are often identified in terms of philosophical movements or leading ideas. They are so identified even by Marxists who are at pains to point out that philosophical ideologies only reflect broader historical or material changes in modes of production.)

(2) Social problems and conflicts over the values needed to solve them are today as pressing as, if not more pressing than, they have ever been in the past.

(3) Academic philosophy of the ivory-tower sort typically shies away from such conflicts; it does not usually attempt to collaborate with those in society who are consciously working to resolve value conflicts. (Dewey proposed this arguement long before the so-called applied turn in philosophy, but I think I can be faithfully Deweyan in saying that the number of applied ethicists today who are involved in social and political activism in areas related to their academic publications would strike Dewey as pitifully minuscule.)

The flow of the argument needs to be interrupted at this point to face two objections. One would say that at least two conclusions are possible at this point: *either* that philosophers ought to return to the old tradition, *or* that activism, in our contemporary division of labor, like so many other things, ought to be left by philosophers to others—in this case, to activists. They might, perhaps, utilize the writings or teachings of applied philosophers—but only if they find them useful. The second objection would ask whether social relevance even *ought* to have been the object of philosophy in the first place. Philosophers at their best have always been clarifiers, not social problem solvers.

Dewey addresses the first objection more or less explicitly, though obliquely:

(4) Academic philosophy, however esoteric it later came to be, was itself part of a social movement in Western higher education responding to conflicts over the value of science and objectivity in dealing with social problems. The demand that philosophers give up their old role as secular preachers and take on the new role of academically respectable researchers was part of a movement that *simultaneously*—though some would say paradoxically or even contradictorily—placed on academics a *service* burden. Finding shelter for freedom of thought behind ivy-covered walls was never, in the public mind, meant to be an escape from social

responsibilities. Dewey's indictment of academic philosophy was aimed at philosophers who had made such an escape.

Dewey's answer to the second of these objections—and the final premise in his argument—must be sought in the general presuppositions of his approach:

(5) Creative intelligence is always a response to problems that block social progress; isolated thinkers may see themselves as working alone, apart from if not in opposition to social conventions. They can be pardoned for this misapprehension, since new insights and creative advances must by definition reflect negatively on ideas taken for granted at a particular moment in a group's history. However, unless such an isolated individual thinker is insanely or irrationally antisocial, his or her new insight is and must be thought of as something that will get the group moving again—hopefully in the right, or at least in a better, direction.

Dewey concludes that philosophy always has been and should now be *reformist, activist, socially meliorative*.[5] (I believe Dewey was astute enough to recognize that this conclusion is debatable and that its supporting argument is at best plausible rather than demonstrative.) Furthermore, he concludes that philosophers claiming to practice "creative intelligence" must do so alongside creatively intelligent people from the broadest possible range of fields, including non-academics faced with real-world problems.

THE ROOTS OF ENVIRONMENTAL ACTIVISM

There is by now a long history of environmental activism. Anna Bramwell traces the roots of environmentalism to scientific communities in the late nineteenth century.[6] Robert Paehlke, citing other historians, pushes the sources back to the middle of the nineteenth century (or even earlier) in the wilderness conservation movement.[7] Paehlke notes that many of the founders of the conservation movement in the United States at the end of the nineteenth century were members of socially prominent families, but their activism was often combined with at least an amateur study of science. Even more scientific was the work of George Perkins Marsh. Especially in *Man and Nature: Physical Geography as Modified by Human Action*,[8] Marsh laid the foundations for the modern concept of ecology as the "study of the interrelationships between organisms [including humans] and environment."[9] Other early conservationists, including John Muir and Aldo Leopold, might be said to be more activist than scholarly, but their works also contributed to the development of the modern science of ecology.[10]

SCIENTIFIC ACTIVISM

The role of scientists in the post-1960s environmental movement in the United States is one of the striking features of that movement. Most people date the modern environmental movement, as a large-scale popular movement, from the publication of marine biologist Rachel Carson's *Silent Spring* in 1962.[11] Paehlke remarks:

> In the decade that followed, hundreds of antipollution organizations sprang up around a wide range of issues concerning the unintended alteration of the chemistry of air, water, land, and food. The principal concern of this new movement was not with the preservation of wilderness or distant forests. Rather, it was with the destruction of the environment within or near agricultural and industrial centers and with the biological underpinnings of human health.[12]

In *Since Silent Spring*,[13] Frank Graham notes how, already with Rachel Carson, the new environmental movement managed to combine scientific expertise with popular crusading. Graham acknowledges that scientific findings from toxicology and epidemiology, and especially from the comprehensive field of ecology, were the foundation of Carson's work. But, Graham says, Carson "knew that her book must persuade as well as inform; it must synthesize scientific fact with the most profound sort of propaganda."[14]

Many biologists and other scientists followed in Carson's footsteps, but two are especially well known—Barry Commoner and Paul Ehrlich. Commoner, in *Science and Survival* and *The Closing Circle*[15] (among other works), launched a relentless attack on the mindless use of chemicals in agriculture and industry that reached epidemic proportions after World War II. First as a scientist, then as a propagandist, and finally as a politician in a futile campaign for the presidency of the United States, Commoner sought to persuade large masses of the public that we are on a collision course with nature—but more important, that a different, more ecologically sound course is possible.

Ehrlich is often lumped with Commoner as an environmental propagandist, but also like Commoner he has strong scientific roots for his activism. Ehrlich's first popular book, *The Population Bomb*,[16] may seem longer on propaganda than on science, but in later popular books as well as in publications directed to audiences of scientists Ehrlich's tone is more scientific—if still not dispassionate.[17]

Commoner and Ehrlich have often disagreed over the question whether chemical pollution or overpopulation is the greater source of environmental degradation. Still another popular book, *The Limits to Growth*, by Dennis and Donella Meadows,[18] took a strong stand on the resource depletion side. *Limits to Growth*, however, blamed the impending crisis its authors foresaw on more than just overpopulation; many other causes of predicted worldwide scarcities are identified in that doomsaying—and controversial[19]—book. The Meadowses claimed that their dire predictions were based on science, but their own approach of world-modeling belongs more to statistics and policy studies than to natural science. Nonetheless, many of the tendencies, if not particular predictions, have been supported by scientists worldwide in the nearly twenty years since *Limits to Growth* appeared; see especially the so-called Brundtland Report, *Our Common Future*.[20]

All this scientific activism has been summed up under the heading, "the visible scientists," by Rae Goodell.[21] Focusing on highly visible scientific activists may, however, give the wrong impression. There is a sizable split between those who are activists and those who oppose activism even among ecologists—not to mention biologists more generally and members of other scientific and technical communities.

For instance, an interesting study of Dutch ecologists has been done by Jacqueline Cramer.[22] In the early 1980s, she interviewed sixty-five freshwater ecologists in terms of both their own environmental activism and their perceptions of the appropriate role for their professional societies. On the latter point, over half of the ecologists interviewed expressed fears about their professional societies becoming too involved with activist groups. As to the ecologists' own personal behavior, Cramer found that, "Only a minority of the academic and governmental researchers (12.5% and 16% respectively) participate in one or (rarely) more environmental groups." An overwhelming majority, seventy-seven percent, do not feel it is their responsibility to participate actively in protecting the environment. One researcher is quoted as speaking for this overwhelming majority: "I am not a keeper of ecosystems; I am a researcher of ecosystems."[23]

AN EXEMPLARY SCIENTIST/ACTIVIST

I want now to cite the example of a famous industrial scientist who did get involved—Russell W. Peterson. His case seems to me illustrative of several points. Peterson, a chemist, came to Delaware in the 1950s to work for the DuPont Company, where he became a high-level manager. Never just a scientist, he soon got involved in a great many activist causes, from homelessness to work with prisoners, as well as environmental activism. One thing led to another, and he got into politics, ending up being elected Governor of the State of Delaware in 1968. In that office, the accomplishment that made him proudest was the passage of a Coastal Zone Act that outlawed any new heavy industry incompatible with protection of the environment along Delaware's side of the Delaware Bay and its small but tourism-oriented Atlantic shoreline. One aim was to prevent the building of a large refinery, an act that evoked criticism of Peterson from people associated with his old employer, DuPont; some of them said then—and still say—that Peterson and his Coastal Zone Act are obstacles to economic growth.[24]

Peterson served only one term as governor, leaving office in 1972, but he went on to important governmental posts in Washington, first as head of the Council on Environmental Quality, then as head of Congress's Office of Technology Assessment. At CEQ, Peterson oversaw the preparation of impressive pro-environment reports,[25] and his tenure was widely praised by environmentalists. After his service at OTA, Peterson in 1979 became president of the National Audubon Society, where again his work was widely respected.

At the Audubon Society, Peterson was responsible for two significant changes. He redoubled efforts in scientific research, and he committed the society to ever-greater activism. The activism led to large increases in membership and, more importantly, to direct public confrontations with the Reagan Administration over its mismanagement of the Environmental Protection Agency. For Peterson, a solid scientific base is necessary for truly effective environmental activism.

Peterson's efforts reveal what may turn out to be a truism about environmental activism. Neither scientific background nor elective office, neither powerful appointive position in government nor the presidency of an old and powerful environmental group with hundreds of thousands of members—none of these without intense lobbying and support by citizen activists could guarantee protection of coastal zones in Delaware or nationally or, most important, keep them protected against endless efforts to change the law in the name of economic progress. As late as 1989, a full twenty years after the beginning of the coastal protection legislative effort in Delaware, Peterson had repeatedly felt the need to continue his appeal to citizen activists. Fortunately, in his view, they have turned out on call each time there is another threat to modify the Coastal Zone Act.[26]

A PROPOSAL FOR PROGRESSIVE ENVIRONMENTAL POLITICKING

Robert Paehlke, in *Environmentalism and the Future of Progressive Politics*,[27] has provided the most sweeping call for environmentalist politics to date on this side of the Atlantic. Paehlke states his thesis early in the book:

> In practice the [environmental] movement has not usually been more than a loose coalition of interest groups. But . . . it can be developed into an ideology able to see the developed economies through the difficult transition from an industrial to a post-industrial society.[28]

How is this transition to take place?

> Environmentalism as an ideology still lacks the mass following that conservatism, liberalism, and socialism in their time attracted . . . [;] to

> attract such a following . . . environmentalists must develop clear and consistent positions on the full range of political and social issues.[29]

Paehlke writes from a Canadian perspective, but he also aims his views at environmental activists in the United States—and implicitly throughout the so-called developed world. He says that, "Neoconservatism has been consistently and deeply hostile to environmental protection in every country in which it has emerged."[30] While he thinks the movement is on the wane, he is very conscious of the appeal of its ideas to large numbers of voters. To serve as a political counterbalance (here Paehlke is referring primarily to the United States), he notes that, contrary to some people's beliefs, "the less advantaged sectors of society . . . *are* concerned about environmental issues, and might become more so if their most pressing economic needs were met." He concludes: "Environmentalism can be creatively associated with the process of achieving social and economic" equity.[31]

What Paehlke proposes as a program for his new progressive/environmental party (which in the United States he says must be a restructured Democratic Party) is a set of six points that would clearly differentiate it from the ideology of neoconservatism. On the other hand, he also says an environmentalist party must both distinguish itself from some ideas of traditional progressives (he lists seven points) and open itself "to selected parts of the contemporary appeal of neoconservatism" in order to "broaden the potential political constituency of an environmentally informed progressivism."[32]

Some of Paehlke's ideas about broadening the base of environmentalism and turning it into a progressive political party are as appealing as they are pragmatic. However, in my opinion, he has the situation reversed: economic equity, the lessening of the power of economic elites, must come first in any progressive coalition if the environment is ever to be protected in the way Paehlke wants. Environmentalist planks must have an important place in a progressive platform, but social justice must be at the center of any movement that will have enough popular appeal to win out at the polls over neoconservatives and their allies with vested economic interests.

Nonetheless, in the end Paehlke is right: what we need is an environmentally alert new progressive coalition. Why? Because real-world ecological problems are so urgent.

It is important, on the matter of urgency, to cite environmentalists first. Lester Brown, editor of the Worldwatch Institute's *State of the World* series, is as good a source as any. He editorializes:

> Even as the [Berlin] wall is falling, there is a growing realization that the threats to our future come less from the ideological differences of nations and more from the environmental degradation of the planet. Each year the forests shrink, the deserts expand, soils erode, the stratospheric ozone layer that protects us from harmful ultraviolet radiation is depleted, the level of

carbon dioxide and other heat-trapping gases in the atmosphere builds, and the number of plant and animal species on earth diminishes.

If any one of these trends continues indefinitely, it will eventually bring an end to civilization as we know it.[33]

The issue of environmental urgency goes beyond environmentalists, however. In 1983, the United Nations General Assembly established a World Commission on Environment and Development, with the charge to set a "global agenda for change" that would "enable the world to achieve environmentally sustainable development." Commissioners came from countries all over the world. Although some were initially skeptical, by the time they had analyzed what must have seemed countless scientific studies, they reached a remarkable consensus. It was published as *Our Common Future* in 1987.[34] A follow-up conference was held in 1992.

A third witness to urgency is *Preserving the Global Environment: The Challenge of Shared Leadership*, a report of the World Resources Institute and Columbia University's American Assembly. It ends on this note:

On this Earth Day 1990, we call attention to the need for immediate international action to reverse trends that threaten the integrity of the global environment. These trends endanger all nations and require collective action and cooperation among all nations in the common interest. Our message is one of urgency. Accountable and courageous leadership in all sectors will be needed to mobilize the necessary effort. If the world community fails to act forcefully in the current decade, the earth's ability to sustain life is at risk.[35]

CONCLUSION

Leaderhip, this report says, is needed "in all sectors." That certainly includes the community of scientists, engineers, and other technically trained experts. It is not enough to say, with the Dutch ecologist quoted earlier: "I am not a keeper of ecosystems; I am a researcher of ecosystems." Similarly, John Lemons, in his retort to the editor of *Environmental Ethics* (see the beginning of this chapter), is surely correct in saying that environmental ethicists must go beyond mere acquaintance with the findings of ecology and the efforts of environmentalists—that "philosophers and environmental professionals [must] work more closely with one another in mutual projects."[36]

So, in the end, John Dewey's view on philosophy generally applies in a special way to environmental ethics: ecological science and environmental ethics (even an applied environmental ethics intended to have an impact) alone, and separately, will not get anything done to save the earth—including humans as part of the ecosystem. Politics is required, a politics involving environmentalists but aimed at developing a broad enough constituency to counteract powerful anti-environmental forces. Whatever a comprehensive science of ecology or an enlightened and activist environmental ethics might offer, political opposition can only be challenged by political means.[37]

Part IV

Progressive Liberalism and Its Opponents

11

A Neo-Marxist Challenge to Progressive Activism

An institutional world . . . is experienced as an objective reality. . . .[Institutions] have coercive power over [the individual], both in themselves, by the sheer force of their facticity, and through the control mechanisms that are usually attached to the most important of them. The objective reality of institutions is not diminished if the individual does not understand their purpose or their mode of operation. He may experience large sectors of the social world as incompatible, perhaps oppressive in their opaqueness, but real nonetheless.

—*Peter Berger and Thomas Luckmann*

If power in capitalist societies is not pluralistic now, then what basis is there for thinking that it can plausibly be pluralistic in the future, especially in the "public interest" sense? Can public-interested members from the different special-interest groups . . . wrest power from the special-interested capitalist class? According to my arguments, such a power transformation is highly improbable.

—*Bernard Gendron*

As with several earlier chapters, I begin here with a quote from Peter Berger (this time with Thomas Luckmann).[1] In this case, his view has links to Marxism, and the quote from Bernard Gendron[2] underscores the Marxist objection to my approach. Jacques Ellul goes further claiming that no one—not philosophers or scientists or engineers or managers or politicians—can exercise mastery over technology as a whole:

> An individual can doubtless seek the soundest attitude to dominate the techniques at his disposal. . . . But the individual's efforts are powerless to resolve in any way the technical problem in its universality; to accomplish this would mean that *all* men adopt the same values.[3]

It is time now to take stock before turning to the more speculative or theoretical part of this essay. To take stock, I would set up a spectrum of assessments, from the most pessimistic to the most optimistic, as to where we stand on technosocial problems today—and, more important, where we can expect to be in the next twenty years or so.

Critics such as Ellul, leaning toward the pessimistic end of the spectrum, would likely say that nothing very promising is going to eventuate from all the activism—including activism by technically trained people—reported in the preceding eight chapters. Family life in our technological world, these critics would say, is, if anything, only going to get worse—especially if we focus (as in chapter 3) on the problems of young people. Those who think, for instance (the pessimists would say), that the situation for black teenagers, whether females having babies before they are ready or males looking for jobs in an increasingly technological work world, is likely soon to improve measurably are simply deluding themselves. And all that hoped-for social-worker activism is not likely to persuade stingy and recalcitrant state legislatures to pour more money into programs for undernourished and neglected children, especially when that can be turned into negative campaign charges about lavishing funds on welfare cheats.

Education, these pessimists would continue, is also only going to get worse. After fifteen or twenty years of alleged reforms, what do we have to show? The schools, by and large and allowing for a few exceptions, are still under the control of the same educational bureaucrats as before. Technological illiteracy and inadequate preparation for real jobs in the real world is likely to get worse rather than better.

Technological domination is sure to continue in medical schools. There, a decreasing pool of medical students will continue to opt for high-technology specialties or biomedical research, and fewer and fewer will go into old-fashioned primary care. (This is still the pessimists speaking.)

Among general problems—that of antidemocratic trends in politics, resulting from ever-increasing dependence on TV advertising but also from advertising-related, high-tech gimmickry and polling—is going to get worse, not better. And the reforms of the reformers, as always, will end up making things worse rather than better.

Under the heading of specific technosocial problems, where we ought to look for scientists' and engineers' and computer scientists' activism to accomplish the most—these pessimists would continue—the picture includes only a small number of bright spots, and the improvements there are the result of broader social forces rather than of activism by members of the technical community. Biotechnology and bioengineering will move ahead as if the naysayers never existed. (Defenders would add that this is a good thing, too; it is clearly good for the beneficiaries of genetic surgery and similar new techniques.) Can we stop data banks and computer and other intrusions into citizens' or workers' privacy? No way, no matter what activists try to do to stop these developments. If the nuclear arms race slows, it will not be because of what activists do (there are not likely to be many activists among nuclear weapons researchers anyway), but because of the end of the Cold War and other historical changes at the international level. Only in the area of environmentalism can we hope for much. And the little achieved there—pockets of wilderness or a few endangered species protected, or areas of shoreline made safe for the middle and upper classes—will hardly slow the pace of technological development more than a few seconds. A pessimistic assessment of scientific and technical activism, in short, would be just as pessimistic as the pessimistic assessment of progressive reforms in broader social arenas.

For my part, I would not move toward the opposite extreme, toward an unqualified optimism on technosocial problems, broad or narrow. I am optimistic enough to say only that I think there is the *possibility* of more progress on these issues than the pessimists predict.

In this and the next two chapters I respond to the pessimists by showing what the basis is for a limited optimism. In this chapter, I take up the challenge of neo-Marxism as formulated by Herbert Marcuse. In chapter 12, I turn to Ellul and similar thinkers, under the heading, "autonomous technology." Then in chapter 13 I turn to the small amount of activism I see addressing the issue the radicals concentrate on: massive technoeconomic power. There I acknowledge that it is unions that have done the most to address issues of economic injustice in a technological world. But I also acknowledge, there and elsewhere, that it may be the case that the only really effective counter-force to massed technoeconomic power would be a coalition of all the movements involved in activism to solve technosocial problems. It may be idealistic to propose such a New Progressive Movement (or party), but the radicals would suggest that perhaps such idealism is our only hope.

MARXISM'S CHALLENGE TO PROGRESSIVE LIBERALISM

At least twice before[4] I attempted to defend a progressive liberal social philosophy of technology against Marxism—only to have my dismissal of that view dismissed as glib[5] or utopian.[6] Albert Borgmann,[7] in spite of being much more careful in analyzing the version of Marxism he chose to attack, has met much the same fate; what he has been accused of, in rebuttal, is a failure to deal with the latest version of Marxism, the one claimed to be the most authentic or the most appropriate for our technological era.[8]

I am not here primarily concerned with the latest iteration of Marxist theory.[9] What I am concerned with is the real-world problems Marxists claim to be addressing—the social problems for which they claim to have the best *practical* (or *praxical*) answer. Like Karl Marx[10] himself, I am not so much interested in philosophical disputes over the best interpretation of the world; I am interested in changing our world for the better. As I have said so many times in this book, I stand with John Dewey and George Herbert Mead[11] as a social meliorist; and I am convinced that a progressive, liberal-democratic politics is more likely to deal effectively with the threats of technology than a Marxist social revolution.

As the Communist world in Eastern Europe comes apart at the seams, it may not seem fair to add one more voice to the chorus of criticisms of Communism. In my view, however, Marxism can still be given a reasonable defense; it can even be defended as the best way to deal with the social ills associated with technology.[12]

WHY MARXISM SEEMS TO OFFER A SOLUTION FOR THE SOCIAL PROBLEMS ASSOCIATED WITH TECHNOLOGY

I do not want here to be accused once again of glibly dismissing Marxist responses to the problems of technology. I take the Marxist response seriously in spite of the end of the Cold War, and I begin here by showing why.

I proposed early on in this book a list of ten types of social problems that beset contemporary high-technology society.[13] The problems range from the nuclear arms race to commercialization of traditional high culture, from ecological catastrophes and genetic engineering to boredom in high-technology jobs and alienation in family life in today's sprawling urban centers. But at the center of my list is growing technoeconomic injustices, and especially the increasing disparity between the haves and the have-nots —whether these are national, between socioeconomic classes in high-technology economies, or international, between developed and supposedly developing nations.

It is this problem that Marx, and Marxists ever since, have focused on. I would go so far as to say that any interpretation of Marx that does not focus primarily on the class struggle between, on one hand, those who control the means of production appropriate to a given stage in the dialectic of history, and, on the other, the exploited workers who actually produce economic wealth is not within the mainstream of Marxist theory as I understand it. I would go further and say, anticipating objections to my view, that any authentic Marxist ought to say that none of the other problems of technological society I list will be solved until the class struggle is resolved world-wide.

Why is this? There would seem to be an obvious link between the economic issue—especially if interpreted in class-struggle terms—and all the other issues: the nuclear economy obviously; industrial and consumption-driven wastes; the temptation of the haves to use high-tech surveillance methods, and perhaps eventually genetic intervention, to keep the exploited have-nots in line or to mold them for particular sorts of work; bribes for workers to induce them to accept hazardous or mind-numbing jobs; worker alienation carrying over into family life, or even leading to its breakdown; schools turned into corporate training grounds without attention to their traditional role of educating responsible citizens; politics turned into media manipulation, frustrating true democracy; the arts no longer critical of society but corporation-dominated. This all-too-familiar litany of contemporary social problems almost always sounds, to defenders of the corporations and of high-tech society, as though it *must* come from left-wing enemies of capitalist society—"fellow-travelers" at worst, or dupes of the Communist line at best.

Several common interpretations of what is going on here need to be dispatched quickly. Students, when they come in contact with Marxist views on the impact of economic power on social problems, often think of it in terms of the exercise of raw economic power. Wealthy individuals, high-level corporate managers, politicians in league with the wealthy and managerial classes, can simply do as they will. If it means profit for them, they can start wars or keep cold wars going indefinitely. (Perhaps they would now say almost indefinitely.) Similarly, critics often take Marxists to be saying that leaders of the capitalist exploiting class act in conscious concert to control education or the media. And, finally, cynics interpret capitalist exploiters as pure and simple greedy men who will do anything, no matter the effects on workers or on the environment, if it means more short-term profits for themselves. (Short-term profits, of course, turn into long-term capital investments, and the cycle goes on.)

None of these interpretations is necessarily or entirely false. No doubt leading capitalists do exercise raw economic power, do sometimes act in collusion in ways that seem to amount to conspiracy (or monopolistic practices), and can be as greedy as anyone else in society. But none of this is the point of the Marxist claim that class divisions pitting capitalists against workers are the root of all social ills in our technological society—or in any previous version of capitalist society. According to Marxist theory (as I am interpreting it here), it is not the individual motives of capitalists, singly or acting in concert, that explain why class-division disparities between capitalists and workers lead inevitably (according to this view) to toxic wastes, hazardous workplaces, and boring high-technology jobs. What makes these social problems insoluble until exploitation ends, according to Marxism so interpreted, is that capitalism is a wholesale ideological superstructure erected on the base or substructure of capitalist-era modes of production. Our entire way of life, all our social relations, not only at work but in the home and everywhere else, are intelligible only in terms of the ideology of capitalism (or, in the present view, techno-capitalism).

Eugene Genovese provides a telling picture of how all of this is supposed to work in his interpretation of life in the slaveholding society of the Old South in the United States, including its accompanying (and legitimating) worldview. The ideology afflicted not only the slaveowners themselves, but their wives, their mores, the law of the land—and even the self-images of non-slaveowning whites, of overseers, as well as of the slaves themselves (however much the slaves later came to see their interests as at odds with the slave economy). In one among many passages (the book must be read in its entirety to get the total picture of a worldview as a seamless—though class-divisive—web), Genovese says:

> This ideology . . . developed in tandem with that self-serving designation of the slaves as a duty and a burden which formed the core of the slaveholders' self-image. Step by step, they reinforced each other as parts of an unfolding proslavery argument that helped mold a special psychology for master as well as for slave. The slaveholders' ideology constituted an authentic world-view in the sense that it developed in accordance with the reality of social relations.
>
> . . . The kind of men and women the slaveholders became, their vision of the slave, and their ultimate traumatic confrontation with the reality of their slaves' consciousness cannot be grasped unless this ideology is treated as an authentic, if disagreeable, manifestation of an increasingly coherent world outlook.[14]

Genovese's marvelously comprehensive account of an earlier capitalist society, where class divisions are obvious, goes into all aspects of the problem—religious legitimations as part of the ideology, and so on. But if his depiction of how economic relations spread out in every direction to become a wholesale ideology seems esoteric and far removed from techno-capitalist ideology, it nonetheless highlights, in a historian's fashion, the substructure/superstructure dialectic. The same thing is done from a social-scientific perspective by Peter Berger and Thomas Luckmann. Their focus is on ideological consciousness and how it comes to have the authoritative character it does throughout a culture:

> Only at this point does it become possible to speak of a social world at all, in the sense of a comprehensive and given reality confronting the individual in a manner analogous to the reality of the natural world. Only in this way, *as* an objective world, can the social formations be transmitted to a new generation. In the early phases of socialization the child is quite incapable of distinguishing between the objectivity of natural phenomena and the objectivity of social formations. . . . All institutions [including the most basic institution of all, language] appear in the same way, as given, unalterable and self-evident.[15]

It should not be thought that such "objectivity" of social institutions, of ideology, ends when the child grows up. As Berger and Luckmann note, one of the most difficult cases for their dialectical theory of social consciousness is that of the alienated intellectual—and especially of the revolutionary intellectual.[16] But far from disproving the wide-ranging impact of reigning ideologies, the case of the revolutionary intellectual actually confirms the theory: it is extraordinarily difficult for *anyone* to break out of an ideology, and, in Berger and Luckmann's view, when one does so, he or she will immediately try to rally a group together and produce a counter-ideology.

Such praxis-oriented revolutionary theorizing has been applied directly to technological society and its problems. The best-known instance is the theories of Herbert Marcuse, especially in *One-Dimensional Man*.[17] For my part, however, I prefer the elaborations of Marcuse's views, in a historical mode, by David Noble.[18] Where Marcuse claims that any opposition to the reigning ideology—for example, in cases of union opposition to hazards in high-technology industrial workplaces—ends up being interpreted as counterproductive, even irrational (according to the "logical" demands of technological "progress"), Noble spells out in relentless detail, and wherever possible in the words of corporate managers, the *total* way in which the ideology of science and technology in the (alleged) service of society came to permeate every aspect of society in twentieth-century America. To speak of solving particular social problems in our science-based economy without changing the overarching ideology, according to Noble (and those who think like him), is, paradoxically, to reinforce rather than undermine the foundations on which the problems rest.

Once again Peter Berger (this time with Brigitte Berger and Hansfried Kellner)[19] can be cited to provide a social-scientific confirmation of this dialectical view. Berger and his colleagues call their method phenomenological, but they intend for their comprehensive account—of how technological production and bureaucracy permeate every aspect of ordinary consciousness in thoroughly "modernized" societies—to be taken to be scientific. They believe that it is impossible to conceive of a modern society without technology and bureaucracy (that is the phenomenological part of their account), but they are equally convinced that empirical studies will confirm the implications of their account. And to deal in any radical way with major social problems such as the boring character of work in highly automated production facilities without changing the overall technoeconomic system would, on their account, seem extremely unlikely. (In fact they think it *is* unlikely in any case.)

What all of this boils down to is a powerful Marxist objection, that reform politics (sometimes disparaged as "mere procedural justice"[20]) cannot get at the roots of technosocial problems without challenging the technoeconomic system. And that system has built-in disparities between exploiting managers and exploited workers, and between high-technology nations and the so-called "developing nations" so often exploited for the raw materials and exotic minerals needed for high-technology production.

Conclusion? If anything is going to be done to deal with technosocial problems, they cannot be dealt with one at a time. They are all interconnected, and *the* fundamental problem is economic. Only a political revolution that eliminates the power of capitalists and quasi-capitalist bureaucrats over the masses of workers offers any real hope of success.

A MEAD/DEWEY-BASED PROGRESSIVE LIBERAL RESPONSE

The first issue to deal with here is the claim, often made by Marxists and other radical critics of techno-capitalism, that the liberal response to technological problems is "merely procedural"—a technical response depending primarily on administrative law and the courts.

Some people who used to call themselves liberals certainly do limit themselves to these approaches; if a technology assessment or environmental impact assessment gets coopted by the corporation(s) responsible for the alleged negative impacts, or if the plaintiffs lose in court, they give up, perhaps saying they will return to the fray at the next legal opportunity.

However, this need not be the only liberal approach, and the progressive social activists I address in this book are much bolder than this in their attacks on technosocial problems. I would, indeed, say that the more progressive, socialist-leaning wing of the liberal movement (in the United States at least) has always been more aggressive than the "merely procedural" objection would suggest. An outstanding example, in my view, is John Dewey. He certainly supported the "polite" legal procedures of the American Civil Liberties Union and the American Association of University Professors. (Dewey was a leading member of both.) And someone could say that his courageous chairing of the Trotsky Inquiry Commission led to no more than procedural justice for Leon Trotsky. But anyone who reads carefully Gary Bullert's *The Politics of John Dewey*[21] and the still more forceful *John Dewey and American Democracy*,[22] by Robert Westbrook, should be convinced rather quickly that Dewey was a dynamic political activist on a wide range of fronts beyond courts of law. Furthermore, as is beginning to be noticed, Dewey related many of the social ills he addressed to the rise of technology-based corporations.[23] My other hero, George Herbert Mead, was as much of an activist on the local scene, in Chicago, as Dewey was on the national scene.[24] I certainly favor procedural reforms, but I am equally convinced that to combat the social evils of technological society much more is needed.

A second prefatory note before my direct answer to the Marxist objection. Both Dewey and Mead were acutely aware of the enormous ideological forces that bear down on an individual who sets out to reform society, and especially a society like ours with the technological propaganda resources at its disposal that Herbert Marcuse has so eloquently pointed out. Two of the best papers at a conference some years ago on Frontiers in American Philosophy were discussions of Mead's social psychology and its bearing on the difficulty of effective social criticism or revolutionary consciousness.[25]

Nonetheless, though Mead and Dewey were aware of the massive power of a technological-ideological superstructure, neither gave in to technological pessimism. In this, I agree with Mead and Dewey, though I believe there is also a great danger in excessive technological optimism. Such optimism as is called for depends on a belief in the effectiveness of Western-style democracy in dealing with technosocial ills. And this belief, further, rests on another belief—hopefully supported by evidence—that even under pressures of technological propaganda and ideology at least a reform-minded activist minority still has the power to make significant political, social, occupational, educational, and cultural changes. Such limited evidence as there is that progressive public interest groups are making some headway against major technosocial evils is collected in Michael McCann's *Taking Reform Seriously: Perspectives on Public Interest Liberalism*.[26] I have also attempted to summarize some of the evidence here, in chapters 3-10.

Given these prefatory notes and beliefs, my principal response to the Marxist objection has three parts.

First, I share Dewey's repugnance to the totalitarian means that have been used by Communist regimes as they have installed their revolutionary forms of government, beginning with the Russian Revolution.[27] As far as I can see, there has been little respect for the civil liberties of ordinary citizens in any Communist country that has sprung up so far. This means that under such regimes, there is little or no opportunity for the sort of public-interest-activist attacks on technosocial evils that I advocate.

Second, there is also, in my opinion, no guarantee that revolution-based Communist regimes, even if they could be established throughout the world and even if they retained their idealistic devotion to the elimination of exploitation, would actually evolve in the direction of a classless society.

Marxists, beginning with Marx himself, have always been convinced (at least as I read them) that once the last vestiges of capitalist exploitation are gone—including the elimination of even the possibility of backsliding into capitalist ways by leaders of the new classless workers' society—all exploitation (almost by definition) would be gone. A new class of workers will grow up in a non-competitive way, free from all the antisocial impulses associated with capitalist competition, highly productive, capable of exercising creative talents of all sorts, and devoted to the common good.

On the other hand, it seems at least conceivable that things would not work out this way at all—that new forms of competition, and eventually of exploitation, would creep back into even the most ideal society.

Christians have faith that the community of saints in the afterlife will not return to the sinful ways of this life. Marxists believe that they have discovered scientific laws of historical development that assure the eventuality of a non-regressive, non-exploitative, classless society in this world. Progressive liberals, like Marxists, prefer to concentrate on this life; but they would rather deal with society as we know it, with antisocial as well as socially responsible people and groups (all to varying degrees). That is why they prefer piecemeal social reform with no guarantee of success.

Third, that brings me to the centerpiece of my progressive-liberal response to the workers' democracy challenge. I believe that progressive social activists are making some headway in terms of combatting the non-economic issues on my list of technosocial ills. (See chapters 3-10, above.) However slowly and haltingly, for example, citizens (and occasionally even their leaders) are responding to the anti-nuclear activists and beginning to see the grave risks the modern world runs because of an escalating arms race. There have also been notable environmental successes in spite of numerous and continuing setbacks. Although opponents of genetic engineering and public spying on citizens have not yet made much headway, at least they are trying to awaken the public to the dangers there. Social workers do not just do direct-service social work but also get involved in activism to deal with problems of drugs, of children, of stressed-out families in urban centers. Activists are working to ensure technological literacy in our public school system. And quite a few organized groups are attempting to deal with technological threats to the democratic process in the West. In short, of the nine issues distinct from technoeconomic injustices, that leaves only one item that is *not* the focus of much social activism—namely, commercialization and corporate domination of the media. In chapter 13, I will focus on such activism as there is on economic issues. That tends to be by labor unions, whose sympathizers have helped pass some laws on occupational health risks, while others are working to create a climate of concern about boredom and mental health problems associated with highly automated workplaces. So there is some hope.

Given these all-too-limited successes—but, even more, given the wide open future possibilities for public interest activism—I now come to the main point of my reply to the Marxist objection. What I maintain is that if we can continue to make progress in solving these technosocial problems, ours will become a significantly better world. And, even more important, in such an improved climate, there is *at least the possibility that newly empowered citizens will move on from past limited successes to attack head-on the overarching issue of technoeconomic injustices.*

CONCLUSION

What I have advocated throughout this book as an antidote to the threat of technological determinism (*à la* Jacques Ellul)—or to the equally serious claim that the threat comes, not from capitalism simply, but from techno-capitalism—is a revival of the spirit of dissent in public interest activist movements. It seems that, to be really effective, activists will have to turn from limited successes on particular technosocial problems to form a nationwide or even worldwide New Progressive Movement. We need, however, to pay heed to the warning of people like David Noble. I would take it as a main theme of Noble's *America by Design* that the reform intentions of the old Progressive Movement were distorted by capitalist owners, corporate managers, leading engineers and scientists in industry, educators, and politicians, turning those intentions from true reform to the maintenance of a capitalism threatened by its own inner contradictions. But we have now read Noble and other critics of our social system, and there is no reason in principle why, in our democratic efforts to improve society, we cannot learn from the past. A New Progressive Movement just might be able to tame technology.

I started my discussion of Marxism by saying that, like Karl Marx, I am more interested in improving technological society than in providing a philosophical understanding of it. In my view, Western democracy—if pursued vigorously in the spirit of dissent of the most progressive social activist groups—is more likely to bring about this happy (but always improvable) state of affairs than any other prescription with which I am familiar. And that includes Marxism in all its forms, even those that have survived the downfall of Communism in Eastern Europe.

12

Autonomous Technology Theorists: A Second Challenge

> *The age of the fulfillment of metaphysics—which we descry when we think through the basic features of Nietzsche's metaphysics—prompts us to consider [to] what extent we find ourselves in the history of Being. It also prompts us to consider—prior to our finding ourselves—the extent to which we must experience history as the release of Being into machination [i.e., into machine technology].*
>
> *—Martin Heidegger*

Where many people nowadays tend simply to dismiss Herbert Marcuse, I chose in chapter 11 to take him seriously—before arguing that his totalizing "one-dimensionality" thesis is challengeable. I want, in this chapter, to set aside for the moment the question whether the only way to challenge the techno-capitalism that gives Marcuse's theory its bite is by way of a New Progressive Movement. Instead, I focus here on other people who tend to agree with Marcuse about technological determinism. I turn, that is, to a consideration of the "autonomous technology" thesis of Jacques Ellul—along with a consideration of the Nietzsche-based view of Martin Heidegger—that the fate of our technological world is necessarily nihilistic.

I find Heidegger's prose hard to fathom, but I think the headnote above from *Nietzsche IV, Nihilism,*[1] sums up at least some of the main points he means to make.

As for Ellul, rather than try to summarize his views, I will let him speak for himself:

> I shall confine myself here to recapitulating the points which seem to me to be essential to a sociological study of the problem [of technology]:
>
> 1. Technique has become the new and specific milieu in which man is required to exist; one which has supplanted the old milieu, viz., that of nature.
>
> 2. This new technical milieu has the following characteristics: it is artificial . . . autonomous with respect to values, ideas, and the state . . . a closed circle . . . not directed to ends . . . an accumulation of means which have established primacy over ends . . . impossible to separate its parts so as to settle any technical problem in isolation.
>
> 3. The development of individual techniques is an ambivalent phenomenon.
>
> 4. Since Technique has become the new milieu, all social phenomena are

situated in it [rather than the other way round]. . . .

5. Technique comprises organizational and psycho-sociological techniques [but] it is useless to hope that the use of techniques of organization will succeed in compensating for the [bad] effects of techniques in general. . . .

6. The ideas, judgments, beliefs, and myths of the man of today have already been essentially modified by his technical milieu.[2]

In chapter 2, I argued against holistic thinking by proposing a thought experiment. Suppose we had solved at least half of the major technosocial problems facing humankind today—problems ranging from nuclear proliferation to cultural upheavals, from technomedia domination of politics to ecological catastrophe. I suspected that the "grand theorists" in philosophy of technology would still be dissatisfied, whereas pragmatists like myself would be overjoyed—though still ready to continue the never-ending struggle for reform. In this chapter, I want to take a somewhat more conciliatory stand. Ellul likes to quote the slogan: "Think globally; act locally." I want to argue that the *real* payoff of these allegedly global thinkers is not global but local. In doing so, I am consciously following in the footsteps of John Dewey in *Reconstruction in Philosophy*.[3]

Instead of addressing Heidegger and Ellul directly here, I turn to their American disciples, Albert Borgmann and Langdon Winner. Only at the end do I comment on Ellul and Heidegger themselves.

LANGDON WINNER: THE MELLOWING OF A TECHNOLOGICAL RADICAL

Winner's *Autonomous Technology*[4] restates one of the major theses in what he calls a "Great Debate" over technology: namely, that it has so far escaped human control. This is, obviously, Ellul's thesis now given a bold new restatement in terms of what Winner calls "technopolitics."

Winner begins with an admirably clear statement of the state-of-the-issue on technology. "Technology," he notes, in past decades "had a very specific, limited, and unproblematic meaning"; today it has become problematic in the extreme, and "it soon becomes clear that in this enlightened age there is almost no middle ground of rational discourse . . . [as] conversations gravitate toward warring polarities and choosing sides."[5] Technology, in short, has become the subject of *the* Great Debate in contemporary culture.

Winner's thesis is stated early: "Ideological presuppositions in radical, conservative, and liberal thought have tended to prevent discussion of . . . technics and politics." Again, "Despite its widely acknowledged importance . . . technology itself has seldom been a primary subject matter for political or social inquiries. . . . Writers who have suggested the elevation of technology-related questions to a more central position . . . have for the most part been politely ignored."[6]

The overall argument structure of *Autonomous Technology* is this. Winner first outlines "several issues centering on the phenomenon of technological change."[7] He then outlines, rather briefly, theories espoused to explain the phenomenon. He includes the theories of Lewis Mumford and Lynn White, Jr., of Max Horkheimer and Theodor Adorno, of Ellul, Heidegger, and William Leiss. All these are found wanting, as theories, in contrast with "technological politics"; what we need, in Winner's view, is to "understand" by radical critique—a variation on Marx's understanding-by-*praxis*.

> The theory of technological politics . . . insists that the *entire structure* of the technological order be the subject of critical inquiry. It is only minimally interested in the questions of 'use' and 'misuse,' finding in such notions an attempt to obfuscate technology's systematic (rather than incidental) effects on the world at large.[8]

One would normally think, Winner's argument continues, that socio-political means for understanding/handling the situation include conservative and liberal, as well as radical political approaches. But, Winner argues, none of these serves as an adequate critique:

> [The] new breed of public-interest scientists, engineers, lawyers, and white-collar activists [represent] a therapy that treats only the symptoms [and] leaves the roots of the problem untouched. . . .
>
> The solution [Don K.] Price offers the new polity is essentially a balancing mechanism, which contains those enfranchised at a high level of knowledgeability and forces them to cooperate with each other. [John K.] Galbraith's cure holds out a virtuous elite within an elite to champion values lost in the new chambers of power. . . .
>
> The Marxist faith in the beneficence of unlimited technological development is betrayed. . . . To the horror of its partisans, it is forced slavishly to obey imperatives left by a system supposedly killed and buried.[9]

Winner concludes this part of his argument: "It can be said that those who best serve the progress of technological politics are those who espouse more traditional political ideologies but are no longer able to make them work."[10]

Winner then comes rapidly to his conclusion, namely, that the only thing that makes sense in a world of technological politics is "epistemological Luddism." This approach "would seldom refer to dismantling any piece of machinery. It would [rather] seek to examine the connections of the human parts of modern social technology"[11]—and undo them where they no longer serve human purposes.

Focusing on these bare bones of the argument of *Autonomous Technology* obscures the fact that Winner's restatement of the thesis of technological politics—an amalgam of themes borrowed mainly from Ellul and Marcuse—is extremely nuanced. His exposition runs to over sixty pages and his argument in support of the view to about twenty-five pages.

Winner's *Autonomous Technology* is an articulate, enlightened, intelligent book, extremely persuasive in its restatement of the theses of Ellul and Marcuse that so many have found unpersuasive heretofore. As I have said before,[12] the question whether it will ultimately be adjudged a wise book is a difficult one to answer. More likely, it seems to me, if Winner's theses are ultimately judged to be wise, they will be attributed to his sources and his book will be credited as the most useful exposition for Americans of those theses.

But Winner's intellectual development did not end in 1977. I want now to turn to his equally articulate later book, *The Whale and The Reactor*.[13]

Winner's notion of "technological politics" had always needed clarification, and Winner has now provided that clarification in one of the papers reprinted as a chapter in *The Whale and The Reactor*: namely, "Techne and Politeia: The Technical Constitution of Society."[14]

In simplest terms, "technological politics," as used in *Autonomous Technology*, had meant that choices of particular "technics" or technologies or technological systems for doing things have political implications. Setting up a particular sort of system for manufacturing a product (and, eventually, for marketing and consuming it) nowadays almost always dictates the presumptively legitimate political relationship between authorities and subjects, between managers and workers—and often ultimately between the managerial classes and the working classes in society. What is especially peculiar about modern technologies, however, is that these political implications are very often completely obscured or hidden by appeal to the demands of technology or scientific rationality. "There is no other way to set up the machinery," respondents to criticism would say, "if we want maximum return on our invested capital"—or "the most efficient use of our production system."

This formulation of the thesis tends to mask its originality, as well as the exact sense in which it is an autonomous technology theory. In "Techne and Politeia," Winner restates his thesis by an appeal to the history of political theory—to the way major political thinkers throughout Western history have dealt with technology. The crucial turning point for Winner is complex. What he does, in effect, is relate the American Revolution to the Industrial Revolution. "The framers of the American Constitution were, by and large, convinced . . . [that] republicanism and capitalism were fully reconciled." But, according to Winner, the situation changed fairly quickly. "There are signs that a desire to shape industrial development to accord with the principles of the republican political tradition continued to interest some Americans well into the 1830s"—but not long thereafter. That is, early in the nineteenth century, political thinkers were still trying to control emerging technology constitutionally. By the end of the nineteenth century, Winner maintains, this was no longer the case. People were so convinced of the blessings that would flow from science that "the form of the technology you adopt does not matter." Winner quotes a *Scientific American* writer in 1896 as saying that it makes no sense to worry any longer about ancient political philosophy concerns—the "empty speculative philosophy of the past."[15]

Winner's version of the autonomous technology thesis is that this new tradition of repudiating traditional political concerns has become entrenched and expanded in the twentieth century. Today, he believes, almost no one thinks of asking what the *political* implications of new technologies might be—and he, along with other radical thinkers, is convinced that a great many technological systems are authoritarian in ways they need not be and in ways that democratic theorists ought to find objectionable.

The Whale and The Reactor includes several trenchant examples of this forgetfulness of politics. None of these examples is stated more eloquently than the one that appears in the title essay at the end of the book:

> Although I had known some of the details of the planning and construction of the Diablo Canyon reactor, I was truly shocked to see it actually sitting near the beach that sunny day in December. As [a] grey whale surfaced [in the distance behind the reactor], it seemed for all the world to be asking, Where have you been?[16]

Winner says his answer had to be that he had "been in far-away places studying the moral and political dilemmas that modern technology involves, never imagining that one of the most pathetic examples was right in [his] hometown."

Immediately, Winner draws his "technopolitical" conclusion:

> From the point of view of civil liberties and political freedom, Diablo Canyon is a prime example of an inherently political technology. Its workings require authoritarian management and extremely tight security. . . . What that means, of course, is that insofar as we have to live with nuclear power, we ourselves [as well as the plant workers] become increasingly well policed.[17]

Winner brings his essay—and the book—to a conclusion with a small anecdote:

> Two years after my epiphany I was invited back to my hometown to give a lecture on technology and the environment. During the talk I argued that while Diablo Canyon was not a very good place for a reactor, it would still be a wonderful spot for a public park . . . [where] parents could take their

children . . . and think back to the time when we finally came to our senses.[18]

Clearly Winner made this proposal tongue-in-cheek. But to me it is also a clear lesson about where Winner's maturation as a thinker has led him. I return to this at the end of the chapter.

ALBERT BORGMANN: A MODEST NEO-HEIDEGGERIAN

Albert Borgmann, in *Technology and the Character of Contemporary Life*[19] is poetic, as I note in a review[20] that begins with these quotes from Borgmann's book:

> The great meal . . .where the guests are thoughtfully invited, the table has been carefully set, where the food is the culmination of tradition, patience, and skill and the presence of the earth's most delectable textures and tastes, where there is an invocation of divinity at the beginning and memorable conversation throughout. . . .
>
> The great run, where one exults in the strength of one's body, in the ease and the length of the stride, where nature speaks powerfully in the hills, the wind, the heat, where one takes endurance to the breaking point, and where one is finally engulfed by the good will of the spectators and the fellow runners. . . .
>
> Like a temple or a holy precinct, the wilderness is encircled and marked off from the ordinary realm of technology. To enter it, we must cross the threshold at the trailhead where we leave the motorized conveniences of our normal lives behind. Once we have entered the wilderness, we take in and measure its space step-by-step. A mountain is not just a pretty backdrop for our eyes or an obstacle to be skirted or overwhelmed by the highway; it is the majestic rise and elevation of the land to which we pay tribute in the exertion of our legs and lungs and in which we share when our gaze can take in the expanse of the land and when we feel the cooler winds that blow about the peaks.[21]

Much of Borgmann's focus is on these "focal things and practices," which partly explains the poetry of his approach. But he is also intent on pointing out that, amidst the clamor of our technological world, there are poetic authors who have highlighted focal things. The quotes above, about the culture of the table and running, are largely borrowed:

> To discover more clearly the currents and features of this, the other and more concealed, American mainstream, I take as witnesses two books where enthusiasm suffuses instruction vigorously, Robert Farrar Capon's *The Supper of the Lamb* and George Sheehan's *Running and Being*. Both are centered on focal events, the great run and the great meal.[22]

Borgmann even claims that he could not have undertaken his project—his phenomenological or "deictic" characterization of the truly important features that can redeem our troubled technological world—if there were not other souls with similar thoughts (and writings) to spur him on and give hope to the project.

But Borgmann is also a philosopher, and his book deserves to be analyzed—even argued with—as well as savored. One of the beauties of the book is that the philosophical argument is presented with as much simplicity and grace as the descriptions of focal things, events, and concerns.

Technology and the Character of Contemporary Life is a tightly-structured philosophical treatise. Borgmann begins the book with a summary of the theories he opposes: "These summaries distinguish a multitude of approaches, but all distinctions fit well one of three essential types: the substantive, the instrumentalist, and the pluralist views of technology." However, Borgmann is modest about the originality of his own theory:

> Clearly, the theory of technology that we seek should avoid the liabilities and embody the virtues of the dominant views. It should emulate the boldness and incisiveness of the substantive version without leaving the character of technology obscure. It should reflect our common intuitions and exhibit the lucidity of the instrumentalist theory while overcoming the latter's superficiality. And it should take account of the manifold empirical evidence that impresses the pluralist investigations and yet be able to uncover an underlying and orienting order in all that diversity.[23]

The theory that Borgmann proposes to meet these exacting demands is his own version of neo-Heideggerianism. He claims to discern a pattern of taking up with reality—the "device paradigm"—which characterizes life in the modern world. (I would paraphrase what Borgmann means by "device paradigm" roughly as the claim that humans, in the modern world, have tended more and more to look for gadgets or devices or systems that will make life easier—at the risk of emptying all "focal" things of their traditional significance.)

Before summarizing the various theories, Borgmann had characterized his mode of philosophizing as derivative from Aristotle as well as Heidegger (for both of whom, despite their differences, he says "there is no sharp dividing line between social science, or perhaps social studies, and philosophy"), yet it is also an approach that takes seriously "the metatheoretical turn" of analytical philosophy. In the end, Borgmann says, he will show, by using it at the beginning, that an analytical approach to philosophy of technology must be an "inconclusive enterprise." Even so, "the present study has to draw on many of the concepts, methods, and insights of mainstream philosophy to obtain a reflective and radical view."[24]

By the end of the book, all this is clarified—perhaps most succinctly in a chapter devoted to "political affirmation" of the possibility of reforming our technological way of dealing with reality:

> These suggestions, drawn from the analysis of technology and the experience of engagement [with focal things], are mere hints, of course. But they shed new light, I believe, on a problem that has become puzzling and untractable within the liberal democratic tradition. They are essentially consonant, however, with the proposals to achieve greater social justice as they have been formulated by the best proponents of that tradition, for example, [John] Rawls, [Lester] Thurow, [and John Kenneth] Galbraith.[25]

That is, Borgmann is "radicalizing" the analytical theory of justice of Rawls and the post-Keynesian economics of Galbraith and Thurow by bringing out the "focal" concerns of a minority within technological culture—including himself, but also such authors as Capon and Sheehan, mentioned earlier. Borgmann is opposed to Marxist radicalism (a version, in his opinion, of instrumentalism, no matter that Marxists claim to oppose it), as well as the radicalism of the right (where, presumably, he would place Ellul—or, at least, Ellulians who would wish to return to a pretechnological golden age).

It is in part three that Borgmann discusses the possibilities of reform. Its main vehicle, Borgmann claims, is *public* "deictic discourse"—the reopening of "the question of the good life," as opposed to continued preoccupation with the consumption of device-procured commodities. Borgmann ends the book, in a chapter on "recovery of the promise of technology," with a nuanced summary of the basis of his hope:

> The focal things and practices that we have considered . . . are not pretechnological, i.e., mere remnants of an earlier culture. Nor are they antitechnological, i.e., practices that defy or reject technology. Rather they unfold their significance in an affirmative and intelligent acceptance of technology. We may call them metatechnological things and practices. As such they provide an enduring counterposition to technology.[26]

How hopeful is Borgmann? I believe it is safe to say that, though he ends the book with an expression of hope that focal concerns will prevail, his worries were serious enough to motivate him to write the book—perhaps as a warning, and at least as a rallying cry for the "concealed" minority who already care more about focal things than about the promise of technology to provide ever more commodities.

CONCLUSION: THE LESSONS I WOULD DRAW

John Dewey, a long time ago, claimed that:

> It has been stated [here] that philosophy grows out of, and in intention is connected with, human affairs. . . . [This] means more than that philosophy *ought* in the future to be connected with the crises and the tensions in the

> conduct of human affairs. For it is held [here] that in effect, if not in profession, the great systems of Western philosophy all have been thus motivated and occupied.[27]

This is unquestionably true of Winner. Although, as I have noted, he provides in *Autonomous Technology* an extensive summary, analysis, and reinterpretation of the theoretical formulations of Ellul and Marcuse (among others), his intention from the outset had been explicitly to do something about the evils of our technological world. In *Autonomous Technology*, the practical focus is on "epistemological Luddism"—not so much, he says, the systematic dismantling of particular machines as the *intellectual* task of bringing us to our collective senses about the hidden political implications of particular technological developments. (Note the plural.) Though an intellectual's task, this is explicitly practical in orientation.

By the time Winner wrote *The Whale and the Reactor*, this practical orientation had become even more marked. Earlier I quoted Winner's whimsical proposal to turn Diablo Canyon into a park where people could reflect on the time we came to our senses, reestablishing democratic political control over technology. Though made (at least partly) in jest, and certainly not practical in the ordinary sense of that term, this proposal clearly has a practical thrust—not just to come to our senses but to establish political control. And not over something as vague as technology in general—or Ellul's "Technique"—but over nuclear technology, possibly starting with that one installation at Diablo Canyon.

The same is true of Borgmann. Returning to "focal things" from our current commitment to technologically produced commodities under the "device paradigm" is fairly vague and abstract, I admit, but the thrust is again practical. Borgmann has clearly found his way through the thickets of Heidegger's prose, and his analytical rephrasing of Heidegger's views seems to me as accurate as any other I have read. But the analysis and reformulation is unquestionably aimed ultimately at culture-criticism, at criticism of our technological culture. And the goal is reform—probably beginning with the reintroduction of particular "focal" practices in particular locales, at least as a start.

With respect to Ellul and Heidegger themselves—to the extent that I can say I understand their motives—the case may not be so clear. But if Borgmann has read Heidegger's intentions aright, that already makes Heidegger more of a moralist than he is sometimes taken to be. And clearly Heidegger did have a healthy opposition to a narrow focus on means and neglect of ends—a view shared with Winner. To a certain extent, this may also be Ellul's main point in his opposition to "Technique." But surely Ellul meant to go further in characterizing technique as a self-augmenting and total system. Even the broader goal of "sociological analysis," as Ellul insists on calling his approach, can however be seen as having a practical purpose: namely, that of a warning, like an Old Testament prophecy, of the evil of our techno-idolatrous ways. In that case, the ultimate goal would be the same as that of the prophets, to get us to repent from our evil ways. (In this case, *if* that is still possible.)

It seems to me clear that Dewey would not have approved of the global part in the slogan, "Think globally; act locally." But at least in his more expansive and open-minded moments, he would have been forced to recognize the local, practical import of even the most global-sounding philosophers of technology. At that point, however, he would invite them to come down from the clouds and urge them to roll up their sleeves and get to work in serious—and concrete—reform efforts.

13

Workers: A Response to Radical Critics

But there are stirrings, a nascent flailing about. Though "Smile" buttons appear, the bearers are deadpan because nobody smiles back. What with the computer and all manner of automation, new heroes and anti-heroes have been added to Walt Whitman's old work anthem. The sound is no longer melodious. The desperation is unquiet. . . .

During my three years of prospecting, I may have, on more occasions than I had imagined, struck gold. I was constantly astonished by the extraordinary dreams of ordinary people. No matter how bewildering the times, no matter how dissembling the official language, those we call ordinary are aware of a sense of personal worth.

—Studs Terkel

Probably the hope of a general economic reform—of an overthrow of the power of capitalist and bureaucratic socialist managers of the sort that Herbert Marcuse envisioned (and despaired of)—will seem today to be the wildest of pipe dreams, given the state of the world in Eastern Europe. That may seem to take much of the steam out of the call for a New Progressive Movement mentioned in chapters 10 and 11, above. However, if my conclusions about autonomous technology theorists in chapter 12 seem at all persuasive, theories about grand and global transformations end up, if they have any concrete ethical import, leading to a concentration on smaller local issues.

I focus on such smaller issues here—unfortunately in a fashion that has caused some confusion among early readers of the material. Several readers have asked why the matter covered here—activism (especially by unions) on specific technoeconomic issues—was not included in part two, above, on general social issues of technological society. My answer to this objection, and my reason for deferring the discussion of these issues until now, is that the real force of the argument made in this chapter can only be explained as a response to the radical critics discussed in chapters 11 and 12. Nothing can be done, the radical critics say, about particular technoeconomic evils unless the system as a whole is radically altered. What I argue here, in contrast, is that specific economic and work-related problems *are* being addressed—and could be addressed even more effectively if the relevant activist groups fully realized what is at stake.

What I focus on here is a consideration of two problems associated with calls for economic justice that seem to show promise in spite of the radical critiques of the neo-Marxists and autonomous technology theorists. What I have in mind are the issues of meaningful work and plant relocations by multinational corporations.[1] I might have focused instead on proposals for democratization either of technological workplaces[2] or of the technological design process;[3] however, those and similar proposals for technoeconomic reform strike me as almost as sweeping as calls for a New Progressive Movement—or, for that matter, almost as sweeping as calls for an international workers' revolution.

JUSTICE AND MEANINGFUL WORK

The theoretical presupposition of this chapter is that issues concerning what are appropriate relationships in the workplace (including whether there will continue to be a workplace after the managers of some multinational corporation decide to move their operations elsewhere) fall within the scope of general theories of justice.[4]

Issues of meaningful work may seem to be not a matter of justice but a matter of beneficence—of benevolent managers choosing to treat workers better in a post-industrial economy than had been the case with owners in the industrial era. (I will later examine a claim that no such benevolence or beneficence can be assumed here.) However, a fairly easy case can be made for treating issues of (alleged) beneficence, along with stronger non-maleficence or no-harm claims about workers' rights to a safe or hazard-free workplace, as matters of justice—even if one admits that there are irreducible principles of non-maleficence and beneficence. Without spelling it out, the case rests on an Aristotle-like assumption that *all* matters of relationships between individuals—except for relations based on friendship—are matters of social justice. In any case, that is what I assume here.

For factual background, I would refer to Studs Terkel's wonderfully insightful interviews in *Working*, cited in the headnote to this chapter. Terkel's subtitle is perfect: "People talk about what they do all day and how they feel about what they do." And a large part of what they feel is by no means positive. Still, Terkel notes some glimmerings of hope, as indicated in the headnote.[5]

A CALL FOR ACTION

I focus here on Edmund Byrne's *Work, Inc.*[6] In simplest terms, it is an appeal to philosophers who believe in social contract theory to revise their thinking in fundamental ways. The most important way, according to Byrne, is for these ethical theorists to take corporations—especially transnational corporations—more seriously in their speculations on the just state than they have heretofore. The reason for this is simple: transnational corporations today exercise *de facto* sovereignty—a sovereignty that always influences, sometimes equals, and often overpowers the sovereignty of nation states.

Easy as it is to state Byrne's thesis, his is by no means a simple book. Its style is cryptic, dense, and allusive. And the argument is so subtle and nuanced that it is not inappropriate to say that the book contains just one long, convoluted argument that extends from cover to cover.

The premises of Byrne's argument are laid out in an introduction. He begins with a paraphrase of a widespread complaint made by people in the labor movement:

> We had a social contract, and now we don't. The social contract has been broken. Government, business, and labor—each had its role and each understood its responsibilities to the others. All three together, cooperating for the betterment of all. That's how it was, but no more.[7]

Byrne follows this immediately with the acknowledgment that this social contract existed for only a short time (especially in the United States)—roughly from the 1930s until the 1970s. And even then, Byrne says, the contract was from the beginning fatally flawed by a basic assumption accepted by all three parties: namely, that the parameters of the contract were national—and this in two senses. There was never any real commitment of the corporations to the local communities in which they operated and from which their workers derived such strength and meaning as they had; and the corporations were becoming increasingly transnational ("multinational" according to more popular usage).

Byrne's conclusions are conveniently set forth in a separate chapter that brings the book to a close. There are three conclusions, which Byrne labels "factual," "hortatory," and "theoretical."

The factual conclusion is the one stated earlier in the paraphrased complaint of union leaders, but it is bolstered in the conclusion by all the interpretations argued for throughout the book.

> The hortatory conclusion [Byrne states] is this: *workers will be able to counterbalance the concentrated power of corporations only to the extent that they and the communities in which they live come to see their interests as intertwined and learn to defend these interests cooperatively.*[8]

> The theoretical conclusion [he adds] is this: *Social and political philosophy will remain irrelevant to a major social and political issue so long as its practitioners do not deal with the fact that corporations are becoming the world's most powerful de facto bearers of sovereignty.*[9]

Byrne had spelled out who the irrelevant social and political theoreticians are in his introduction, but his primary target is John Rawls. Byrne views Rawls as a *liberal* defending the claim that the public sector has a responsibility to take care of people's (including workers' and their families') basic needs,[10] and he sees Rawls' opponents (e.g., Robert Nozick) as *libertarian* neo-conservatives with their emphasis on the efficacy of individual initiative.

Throughout the book Byrne uses as his means of arriving at his conclusions the method of demythologizing. What he claims to be doing is slaying "dragons that guard the gates of the status quo"—namely, legal assumptions about corporate personhood and eminent domain, or about private property and the commodification of goods; management ideas about employees as autonomous individuals rather than citizens with roots in local communities, plus the management ideology of "profits without payrolls" by way of robots and automation; and, finally, ideologies of progress and competition.

I will here look at three examples of Byrne's demythologizing. The first is concerned with the obligation or right to work, the second with claims about "meaningful work," and the third with obligations of justice in plant relocations.

In part one, "Worker and Community," Byrne deals with three issues: the obligation to work, the work ethic, and responsibility for people who are unemployed. Under the first heading, after reviewing the opinions of philosophers ancient and contemporary on the issue of forced labor, Byrne concludes that "freedom has come to be more highly valued than work . . . [so that] a well-informed representative of workers [Byrne's point of view throughout] would want to proceed with caution before endorsing a social contract in which work is made obligatory."[11]

On the work ethic, Byrne defends a somewhat controversial view about a possible "contractarian basis for [an] obligation [to work] in a just society." He does so by defending four theses, namely that:

(i) not all human beings would recognize or agree to an obligation to work (largely an examination of Johan Huizinga's reading of history in *Homo Ludens: A Study of the Play Element in Culture*);[12]

(ii) not all *rational* human beings would recognize or agree to an obligation to work (people throughout history whom one would not want to accuse of an adolescent predilection for play over work—for example, clerical academics—are cited as evidence);

(iii) not all rational, *responsible* persons would recognize or agree to an obligation to work (here Byrne cites management rules: an ultimate rule, that whenever possible people are to be replaced by machines, and an interim rule that says to use the work ethic to get as much work as possible out of workers in the meantime); and

(iv) not all rational, responsible, *knowledgeable* persons would recognize or agree to an obligation to work. In defending this fourth thesis, Byrne arrives at his all-too-obvious conclusion: that few people value work for its own sake. Or, stated more directly, most people value work only as a means to some other end.

On responsibility for the unemployed, Byrne acknowledges that "a society's welfare benefits may be influenced by presumptions about work obligations," but "nonetheless one's involvement in the workforce does not guarantee eligibility for benefits." About this state of affairs Byrne's indignation shows through:

> We are all losers if we continue to acquiesce in a public policy that for all practical purposes abandons displaced workers like tools no longer needed. We do not cut off benefits to veterans of yesterday's wars just because they served with now obsolete means of destruction. Still less should workers be forgotten simply because they served with now obsolete means of production.[13]

Byrne describes "meaningful work" as a "seductive" notion. As a general proposition, he says, "The more people expect their work to be meaningful, the more they seem to challenge employers' claims to control over the work relationship."[14] And Byrne raises four objections to the expectation of meaningful work: (1) Job satisfaction is not a sufficient reason for keeping a job, and the absence of job satisfaction is rarely a sufficient reason for leaving one. (2) Meaninglessness is not peculiar to disappearing low-skill jobs, and meaningfulness is often missing in new high-skill jobs. (3) In any case, whether a job is viewed as meaningless or not, it is always subject to termination. And (4), no matter how well-intentioned the "meaningful work" movement is, it is peculiarly vulnerable to manipulation by management:

> Under such labels as job enrichment, [quality of work life], and cooperation, [employers] are luring even unionized employees out of deskilled niches inherited from the past into purportedly more complex and challenging assignments. Workers in their turn are expected to respond to this recognition of their potential with deepest gratitude. But gratitude is not the most common response. As these experiments in [meaningful work] are carried out at the workplace (rather than in scholars' skulls) they frequently involve more stress and less compensation.[15]

This may seem to be a pessimistic conclusion, considering the inherent appeal of the meaningful work ideal, and Byrne ends his discussion on an appropriately ambivalent note: "Employers are to be encouraged to provide opportunities for the exercise of creative potential. But people must remain free to decide for themselves how they personally want to go about exercising their own creativity."[16]

JUSTICE AND PLANT RELOCATIONS

Byrne slays his most important dragons and comes to his most important conclusions in part three, "Corporation and Community." But earlier in the book he had already done some heavy demythologizing:

> Plant closings are commonly defended as a matter of business necessity. Many labor-intensive plants have been closed in recent years . . . especially in . . . the so-called rust belt. Why is this the case? Some blame rising labor costs. . . . Others, including [union] experts . . . , prefer to blame "the importance of technological innovation as a means of [meeting] competition." The pressure of competition may generate a desire to innovate. But it may also inspire a company to find an environment in which "cheap labor" is available . . . [or it] may be an opportunity to "get out from under" a union.[17]

Under the heading of plant closings, one dragon Byrne attempts to slay is new laws and legal interpretations that try to restrain the property rights of corporations. But, he says, the corporations display a remarkable immunity to these efforts: "Exemplifying this immunity is the fact that corporate restructuring often undercuts the [National Labor Relations Board's] distinction between partial and total closings, thereby exempting the 'restructuring' employer from notifying and negotiating with its 'lame duck' employees."[18] And he goes on to cite the example of U.S. Steel, transformed into a division of USX, shutting down its mills in Youngstown, Ohio, in 1979.

Later Byrne says:

> Judith Lichtenberg is certainly correct in saying that "the company's ownership of the factory cannot settle the issue of its responsibility in plant closings." But, as we have seen, ownership is not necessarily coextensive with control, and either may change about as quickly as the price of a stock on the trading board. So a narrowly focused insistence on advance notice and transitional benefits already concedes the characterization of a corporation as a commodity and leaves communities in the position of beggars who, as has oft been noted, cannot be choosers.[19]

After which Byrne launches into his last and most powerful argument:

> It is essential that communities . . . be in a position to be choosers. A community being, by my definition, a geographically localized complex of legitimate interests (abstractly) and (concretely) human beings who assign these interests moral priority, the task before us is to tie the community thus understood to a plant or facility which a corporation owns or controls.[20]

Byrne can accuse Lichtenberg[21] of a narrow focus on legalistic definitions, but we should be clear what his opposed focus is on—namely, a broad political restructuring that would give back to communities the power (did they ever really have it?) to negotiate a social contract on an equal footing with multinational corporations. Here we should recall Byrne's overall hortatory conclusion at the end of the book, that workers need to mobilize their power, in local communities, and *defend their interests cooperatively*. Surely Byrne recognizes that this will be seen, at least by managers (and members of what can justly be called the managerial classes), as a call to class struggle—of workers and their communities not only against the owners of corporations but against the whole social, political, and legal system that supports them, and ultimately against the ruling ideology of capitalist society.

I point this out not to disparage Byrne's book. Far from it. But it seems clear to me that Byrne's hortatory conclusion demands far more—in the way of political savvy, political activism, even political *power*—than his final theoretical conclusion does. All the theoretical conclusion requires is that political philosophers be more realistic. But then, if political philosophers became more realistic, they might see the need to go beyond theory to calls for restructuring political power relationships.

CONCLUSION

Is there anything in Terkel's anecdotal evidence in *Working* to suggest that ordinary citizens in the United States would be likely to rally around those who call for an economic restructuring of society? For that matter, does Byrne provide such evidence, or is there any to be found in his voluminous sources?

I would suggest that there is little evidence of a broad social movement in support of greater technoeconomic justice in the United States today. Terkel interviews one union organizer who says:

> [The young workers'] idea is not to run the plant. I don't think they'd know what to do with it. They don't want to tell the company what to do, but simply have something to say about what *they're* going to do. They just want to be treated with dignity.[22]

Terkel quotes another union organizer as saying that "even conservative workers are militant in shops," but the effect of this is undercut when the organizer adds that most of the workers in his industry are black or Hispanic women and extremely vulnerable to layoffs. Even worse, he predicts the eventual elimination of that particular industry, shoemaking, in the United States. Nor is Byrne's evidence, or that of his sources, any more persuasive. Griping about work is clearly pandemic in the United States, and a great many of the gripes seem justified. But this has not so far translated into any mass movement flocking to heed Byrne's hortatory conclusion—namely, that workers and their communities should act collectively and consciously to counter the excessive power of the corporations.

This bleak picture, however, should not be the last word. Earlier I cited Michael McCann as claiming that there are "legions of intellectuals"—people like Byrne—"committed to progressive economic and social policy formulation."[23] And surely Byrne wrote his book with the hope that workers can be motivated to combine their efforts to fight against the power of the corporations. Whether this hope is utopian or not, there is every bit as much evidence that workers, and their communities, *could* organize as there is that they are not currently doing so—or that they are not doing so effectively. That workers, especially unionized workers and their allies in Congress, still have power to bring about some changes is evident in occupational safety and other regulations that have been passed in recent decades. What an optimist can still hope for is that limited successes can lead to larger and larger demands, so that in the end, through a progressive coalition, some closer approximation to justice can be achieved.

14

Conservatives

Following [Alexis de] Tocqueville, whose Democracy in America *is a classic study of the ideal-typical aspects of aristocratic as against democratic social structures, I should like to make a rather broad generalization: whenever the ideals of equality and democratic individualism (Quakerism) are stressed in a society, ambitious men and natural aristocrats will come to the fore through a series of petty power struggles; they will of necessity have small rather than large ambitions involving power and success rather than great accomplishments and fame; at the same time, individuals in inherited positions will have, of necessity at first and normatively later, retreated from authority and ambition to rest on their pecuniary privileges. . . . When, on the contrary, a society is marked by the ideals of hierarchy, class authority, and aristocratic social cohesion (Puritanism), ambitious men, both natural and sociological aristocrats, men of inherited position and those of achieved position, will tend to be driven to accomplishment and fame and be less likely to rest on power or privilege alone.*

—*E. Digby Baltzell*

It would be a fruitless task to attempt to summarize the marvelously textured and subtle interweaving of data and analysis that make up *Puritan Boston and Quaker Philadelphia,*[1] E. Digby Baltzell's attack on the progressive activism of the late 1960s and early 1970s. The book, part history, part sociological theorizing, uses Max Weber's ideal-types approach in a very skillful way. But it is also in very large part normative theorizing—or special pleading, even preaching, if you will—for the values of tradition, family, and a sense of noblesse oblige, and against antinomian values of the "inner light" that lead to rule by managers and technocrats at best, by demagogues and bureaucrats at worst.

Baltzell maintains that "from colonial times to the present, Philadelphia has been led by a far higher proportion of auslanders, [non-upper-class] outsiders, and cut-flowers [one-generation leaders] than Boston has been."[2] Baltzell thinks "this has been largely the result of the very different upper classes in the two cities"—a difference which he then traces, through the remainder of the book, to the contrasting "Protestant ethics" of the two cultures, the one Puritan-authoritarian, the other Quaker-democratic. In a contrast that is revealing, Baltzell says, "I believe it is the myths men live by, rather than either their genes or their social milieus, that determine individual fame or failure."[3]

Baltzell is, thus, an unflinching idealist in the Marxist sense: ideas, especially religious ideals, rule history rather than the reverse.

Baltzell is here defending one brand of conservatism, and it is important to note that. What he says he favors is "deference democracy," not some more extreme form of traditionalist aristocracy. His view is also classified by what he opposes: "defiant democracy," the opposition to authority and tradition that he saw running rampant in the 1960s and early 1970s.

I use Baltzell here to emphasize that there are many brands of conservatism, only some of which embrace technology warmly. In my view, there is no single underlying philosophical system that undergirds conservative political theory. Carl Cohen, in an ideal-type categorization of fascist conservatism, lists among its advocates Plato, Niccolo Machiavelli, G. W. F. Hegel, Thomas Carlyle, and Friedrich Nietzsche—but also Adolf Hitler, Benito Mussolini, and Giovanni Gentile.[4] Russell Kirk, much closer to Baltzell, relies on British conservative sources, from Edmund Burke to Christopher Dawson or C. S. Lewis, along with their American admirers, from John Adams to John C. Calhoun to Robert Nisbet.[5] Other accounts likewise stress fundamental rifts among conservatives, underscoring the split between Continental radical conservatism—for example, of the French type—and the more moderate British school.[6] *The Encyclopedia of Philosophy*, finally, mentions a "skeptical conservatism," best represented by Michael Oakeshott, which is so anti-ideological that it is skeptical even of conservatism itself as a political creed or partisan platform.[7]

NATURAL LAW CONSERVATISM

In spite of this diversity, there is at least one conservative philosophical system that might serve as an exemplary support for a theory of political conservatism—namely, natural law theory. Natural law theory, furthermore, accounts for many of the main tenets of conservatism as outlined by Cohen: its defense (advocates would say recognition) of inequality, its "organic" or natural character, its moral base, and its claim that true freedom is exercised only under lawful authority. Advocates of a natural law version of conservatism tend to see sources other than technology as the root of contemporary evils, but recent popes and Roman Catholic officials have offered natural law prescriptions to cure the ills of technological society.

I refer here to three versions of a natural law approach. The first is obvious; it is that of Thomas Aquinas—after seven centuries, still the simplest and most clearly articulated version. This view has been updated and translated into an idiom compatible with contemporary analytical discussions of the nature of law, justice, and morality by John Finnis, for example, in *Natural Law and Natural Rights*.[8] A third version, in the modern idiom but less neat about fine distinctions and more concerned with application to the cure of ills of the modern world, is that of John Courtney Murray, in *We Hold These Truths*.[9]

In presenting Aquinas' natural law theory, many interpreters turn immediately to his so-called "treatise on law."[10] This may be a mistake since it can give rise to misunderstandings. Designed as a textbook to make up for the disorganized and haphazard way in which theology was taught at his time, Aquinas' *Summa theologiae*, in which the treatise appears, is nothing if not organized and structured—to the point that virtually nothing is repeated and virtually every article, every subsection of every question, must be interpreted in light of the systematic whole. (We might prefer today to call it an encyclopedia or handbook rather than a textbook.) The basic plan of the *Summa* is simple: it is an account of God as creator and of humans as coming forth from God and returning to God—i.e., the plan is entirely theological, not philosophical in the fashion of Aristotle or of modern philosophers. But Aquinas was not an anti-philosophical theologian, as were many of his contemporaries; it was his view that revealed truth does not change natural truths—and so he felt free to intersperse within his theological synthesis the best of the natural knowledge available at the time. For him that meant mostly Aristotelian philosophy, and he would have maintained there was nothing in the *Summa* inconsistent with Aristotle's *Politics*.

The first part of part two of the *Summa* deals mainly with the endowments God gave to human creatures to help them negotiate a return to him: awareness of the goal, a free will to pursue it, emotions, habits and virtues (Aquinas also treats their opposites, sin and vice), along with certain superadded divine gifts. (Human constitution in terms of knowledge had been treated earlier in the *Summa*.) The virtues or good habits are treated as "internal" aids, as "inner" modifications or actualizations of human capacities to know, to love and hate, to desire and enjoy, to hope and despair, to fear and become angry (all reasonably, if the acts are virtuous). Law is treated as an "external" aid, as the "outer" presentation (Aquinas was an intellectualist, not a voluntarist, in legal matters) of things to be done and avoided in the pursuit of good. For Aquinas, this theological presentation in no way distorts the natural truth of the matter; had God not given divine revelation, humans would still have pursued a goal of happiness, would need virtuous habits to do good and avoid evil, and would need the "external" or "outer" presentation of things to be done or avoided.

One of the misunderstandings Aquinas' way of discussing law in the *Summa* has occasioned is an overemphasis on this "external" aspect: the dictates of natural law are sometimes treated as if they were a set of propositions—a misunderstanding prompted in part by Aquinas' own practice of equating the basic precepts of the natural law with the biblical Ten Commandments. In fact, what Aquinas claims is that humans spontaneously recognize as good (and their opposites as evil) all those things toward which they have a "natural inclination." The first inclination is simply to do good and avoid evil; others include an inclination to self-preservation, to sex and the rearing of children, to the avoiding of ignorance and to living peaceably with neighbors. Having an inclination to something, recognizing it spontaneously as a good, does not—for Aquinas any more than for Aristotle before him—mean that one will necessarily follow the inclination; some individuals may choose evil—for example, killing a parent—but they will recognize what they have done as evil. The point of both ethics and moral training in the Aristotelian tradition is to help in achieving the ends spontaneously recognized as good.

Among these natural inclinations, Aquinas recognizes a hierarchy. While some inclinations are spontaneously recognized by anyone who has "the use of reason," others will require investigation. And this brings up another misunderstanding occasioned by Aquinas' mode of presentation in the *Summa*: no philosophical or psychological or other expert knowledge is required to recognize natural goods—they are spontaneously recognized by everyone with minimal intelligence, by everyone of a certain age unimpaired by gross mental incompetence, by the proverbial "poet and peasant" as well as by the philosopher. Natural law moralists over the centuries, following the lead of Aquinas, have reduced this insight to a formula of primary, secondary, and lower levels of natural law precepts down to the level of "positive" or civil law. While some such discussions ended up in so-called "casuistic quibbling," the insight is important. In cases of conflict (real or apparent) between natural inclinations, the higher inclination prevails. This order, however, can be inverted on occasion, where it is reasonable. For instance, self-preservation can override living peaceably with neighbors when one is under unlawful attack, and living in society can override self-preservation when society is under attack.

This hierarchy among reasonable inclinations allows Aquinas to deal with two major difficulties for natural law theory: namely, whether it is the same for all people in all cultures, and whether it can change over time. Aquinas, without the support of modern anthropology, claimed that at least the basic inclinations are the same in all cultures; neither do the basic precepts change over time—though natural law inclinations at a lower level of generality, as one gets down to particular cases requiring the intervention of experts, do involve diversity and change. As an example, Aquinas borrows Julius Caesar's prejudiced assertion that the "barbarian" Germanic tribes on the frontiers of the Roman empire were "natural thieves." Presumably what this means is that, according to Aquinas, the Germans would consider it generally wrong to steal, but not from the Roman invaders. Aquinas even claims that the customs of certain cultures—their "depraved customs and corrupt habits"—can entirely block the recognition of "secondary precepts."

Aquinas goes even further than this. Somewhat contradictorily, considering his claim that divine intervention does not change natural law, he explains away biblical cases of ravaging enemy territory, of multiple wives, of God commanding adultery or the killing of an innocent child. It is this unfortunate tendency to make unreasonable exceptions that has sometimes given natural law theory a bad name.

John Finnis, attempting to bring natural law theory up to date, to make it palatable to a generation of legal philosophers raised on H. L. A. Hart or Ronald Dworkin, proposes that we involve ourselves in a practical reflection or private meditation, where "the proper form of discourse is '. . . is a good, in itself, don't you think?'"[11] On this basis he thinks any "practically reasonable" person will come up with seven basic values: life, knowledge, play, aesthetic experience, sociability, practical reasonableness, and religion.

To these values, using the same method, Finnis adds nine ways in which the first principles of natural law, his set of irreducible basic values, can be applied to the building up of a moral system in the modern sense: they must be embodied in a rational plan of life (in this his approach is like that of John Rawls); no basic value can be discounted or exaggerated (*contra* Rawls); morality demands impartial recognition of the right of all humans to pursue the basic values (the "golden rule"); particular projects are to be treated with detachment, not accorded the importance of basic values; yet (fifth) one should not abandon commitments lightly; actions should be judged by their fitness for given purposes (Finnis's limited acknowledgment of the moral importance of consequences or utilitarianism); every basic value, so far as possible, should be respected in every act (intended as his version of "the end does not justify the means" or Kant's "always treat humans as an end and never as a means"); always favor the common good; and (ninth and last) always act in accord with conscience (even if others judge it to be erroneous).

Whether Finnis's opponents—Kantians, or neo-Kantians like Rawls, or utilitarians—will accept this approach as transforming natural law theory into an acceptably modern, even if still debatable, theory of morality is certainly questionable. Kantians are still likely to object to the controvertible character of Finnis's "practical reasonableness," while utilitarians would find the approach too inflexible (where good consequences, in certain circumstances, would suggest the overriding of one of the basic values). Nonetheless, Finnis has gone a long way to "dress up" natural law theory—so that natural law should not be too easily or quickly dismissed by contemporary ethical theorists.

In passing, it should be noted that more traditionalist natural law theorists would also likely quibble with Finnis. He is so convinced of the irreducible character of his seven basic values that he *seems* committed to eliminating the principle of hierarchy—the adjudication of conflicts among basic values when these appear. Perhaps, however,

Finnis would interpret his nine practical or moral-methodology principles in such a way as to allow a hierarchy between basic values and particular projects thereby derived from them. Whether that would satisfy natural law traditionalists is a question. Finnis, in any case, could certainly argue that, for instance, a conflict between self-preservation and the demands of society will only arise at the level of "particular projects."

For my purpose in this book, both of these accounts remain too abstract and general. Assuming that conservatives, in criticizing modern bureaucracy and managerialism (T. S. Eliot or Kirk) or the social ills attendant on the widespread acceptance of a "Quaker ethic" (Baltzell), might ground their critiques in natural law theory, still they would need some less abstract application of the theory to modern life. Fortunately, there is one well known application. It is the critique of post-World-War-II America by John Courtney Murray in *We Hold These Truths*. In Roman Catholic circles of his day, Murray was widely viewed as a liberal rather than a conservative, but he was also—as will be seen—an arch-foe of secular liberals and Marxists and would be viewed by them (as would any advocate of natural law) as a spokesperson for a fundamentally non-liberal, non-radical view. Since one of the views Murray explicitly argues against is nineteenth-century liberal individualism—his argument is that it is insufficiently "organic"—this moves him toward Cohen's ideal-type conservatism, mentioned above. Murray, as a convinced Roman Catholic, also believed in transcendent truth and responsible freedom under lawful authority, so he can be taken as a representative spokesperson for the conservative natural law tradition as applied to practical affairs.

Murray's *We Hold These Truths*, a synthesis of numerous articles written in the Cold War 1950s, is a polemical endorsement of natural law theory as relevant to the problems of pluralism in the modern world, especially in the United States. For Murray, the embodiment of a natural-law-based political system would not be a Catholic state like Franco's Spain or a medieval theocracy; it would be a secular democracy. And only the theory of natural law, Murray thought, could undergird the American experience of pluralism unified into a single nation:

> I take it that the political substance of democracy consists in the admission of an order of rights antecedent to the state, the political form of society. These are the rights of the person, the family, the church, the associations men form for economic, cultural, social, and religious ends.[12]

What is the guarantee of these rights? For Murray it must be some transcendent truth, and in particular the basic tenets of natural law:

> There are truths; and if they are not held, assented to, consented to, worked into the texture of institutions, there can be no hope of founding a

true City, in which men may dwell in dignity, peace, unity, justice, well-being, freedom.[13]

Murray's way of defending this thesis, which seems to fly in the face of traditional interpretations of the American experience, is to contrast natural law theory with other "metaphysical decisions" people might make in attempting to ground a just society in philosophical theory:

> The doctrine of natural law can claim to offer all that is good and valid in competing systems, at the same time that it avoids all that is weak and false in them. Its concern for the rights of the individual human person is no less than that shown in the school of individualist liberalism with its "law of nature" theory of rights, at the same time that its sense of the organic character of the community, as the flowering in ascending forms of sociality of the social nature of man, is far greater and more realistic. It can match Marxism in its concern for man as worker and for the just organization of economic society, at the same time that it forbids the absorption of man in matter and its determinisms. Finally, it does not have to bow to the new rationalism in regard of a sense of history and progress, the emerging potentialities of human nature, the value of experience in settling the forms of social life, the relative primacy in certain respects of the empirical fact over the preconceived theory; at the same time it does not succumb to the doctrinaire relativism, or to the narrowing of the object of human intelligence, that cripple at their root the high aspirations of evolutionary scientific humanism. In a word, the doctrine of natural law offers a more profound metaphysics, a more integral humanism, a fuller rationality, a more complete philosophy of man in his nature and history.[14]

More important, for Murray, it offers a better grounding for democratic pluralism as found in the United States than the philosophical systems generally invoked as a ground—whether individualist liberalism or "scientific humanism"—and something better than democracy's chief competitor at the time, Communism.

If we turn to problems of a technological society, Murray takes up one of them specifically and in some detail—modern warfare. Although Murray grants that war has been fundamentally altered by nuclear weapons (as well as chemical and biological weapons), he ends up espousing a modified version of the traditional natural law theory of the "just war." Another issue Murray addresses is the media. Still another is this: "Society becomes barbarian when technology assumes an autonomous existence and embarks on a course of unlimited self-exploitation without purposeful guidance from the higher disciplines of politics and morals"—where his example is the space race, especially President Kennedy's moon landing project.[15] Murray is also, to some extent, aware of and concerned about the technocorporate state, "that most intricate and powerful combination of science, technology, and business."[16] Finally, with respect to humanity, technology, and nature:

> We have come to disbelieve the cardinal tenet of modernity which

> regarded every advance in man's domination over nature—that is, every new accumulation of power—as necessarily liberating.[17]

We "post-modern" Americans, Murray says, have come to recognize "the horrid vision of man, master of nature but not of himself."

In *We Hold These Truths*, Murray does not offer many more practical solutions to the urgent problems of contemporary society than these platitudes. But Murray was also a man of action, and a golden opportunity opened up for him when Pope John XXIII called into being the Second Vatican Council. It is generally agreed that Murray directly influenced that body's "Declaration on Religious Freedom" and had a strong influence as well on Vatican II's "Pastoral Constitution on the Church in the Modern World"—and thus indirectly on a number of post-Vatican II documents of popes and other Roman Catholic officials.[18]

NATURAL LAW PRESCRIPTIONS FOR A GOOD TECHNOLOGICAL SOCIETY

"The Church in the Modern World," at the time it appeared, was viewed as a liberal, even a radical, document. It spoke of "profoundly changed" human conditions, even of "revolutionary" new insights, from the social and behavioral sciences, into the nature of "human self-knowledge."

> Today's spiritual agitation and the changing conditions of life are part of a broader and deeper revolution. As a result of the latter, intellectual formation is ever increasingly based on the mathematical and natural sciences and on those dealing with man himself, while in the practical order the technology which stems from these sciences takes on mounting importance.[19]

The council fathers go on to note how industrialization and urbanization, together with such other factors as the mass media, have transformed modern society.

Yet "The Church in the Modern world" is also a profoundly conservative document. Its basic aim is "to carry forward the work of Christ"—now not just to believers but "to the whole of humanity." When it later comes to practical problems, such as new versions of agnosticism related to science and technology, the solution offered is this: to harmonize "the proper development of culture" in science and technology with the message of the Gospel. Again, when dealing with the population problem, the document does so in the context of the sacredness of marriage and the family—and ultimately with an unchanged appeal to natural law:

> The moral aspect of any [birth control] procedure does not depend solely on sincere intentions or on an evaluation of motives. It must be determined by objective standards. These, based on the nature of the human person and

> his acts, preserve the full sense of mutual self-giving and human procreation in the context of true love. Such a goal cannot be achieved unless the virtue of conjugal chastity is sincerely practiced. Relying on these principles, sons of the Church may not undertake methods of regulating procreation which are found blameworthy by the teaching authority of the Church.[20]

Post-Vatican II documents of the Roman Catholic Church have shown this same ambivalence: an openness to new social problems, but a tendency to offer solutions that are quite traditional. Pope Paul VI, in "A Call to Action: Letter on the Eightieth Anniversary of *Rerum Novarum*,"[21] sums up many of these "new social problems," together with a range of traditional solutions.

The problems listed include urbanization, with its high-technology specialized unemployment, consumption-oriented economy, and anomic or alienated living conditions; special problems of youth and women; problems of workers and especially of the poor, handicapped, and maladjusted; racial discrimination; migrant workers and unemployment problems generally; mass communications, including their potentially dangerous impact on "the common heritage of values"; and, finally, problems of pollution and the environment related to technological development.

The solutions offered by advocates of natural law theory (not necessarily by Pope Paul VI in this document, but more often by the popes than by other natural law thinkers) have ranged from a greater emphasis on a generalist education to condemnations of materialism to the "principle of subsidiarity" (or workers' guilds), to condemnations of discrimination (while also, in that context, emphasizing "women's rightful place"), to use of the media for proclaiming the Gospel—and including the continued prohibition of "unnatural" birth control (in Pope Paul VI's *Humanae Vitae* and a letter of American Roman Catholic bishops supporting it, as well as in "The Church and the Modern World").

Some have thought they found more liberal solutions for other contemporary social problems—e.g., modern warfare or economic injustice—in earlier papal documents: especially Pope John XXIII's "Christianity and Social Progress" (*Mater et Magistra*) and "Peace on Earth" (*Pacem in Terris*). However, *Pacem in Terris* represents, at root, nothing more radical than John Courtney Murray's updating of "just war" theory already referred to; and while *Mater et Magistra* did condemn the evils of an unfettered *laissez-faire* capitalism, it also contained all the older condemnations of Communism.

In short, while the popes in recent decades have been more willing than other natural law thinkers to confront the new social problems of a technological culture—and Vatican II followed this example—it should be admitted by most impartial observers that the reforms advocated have not strayed far from the traditional line. And this despite the fact that, in traditional natural law language, the differing conditions in different cultures and historical periods ought to allow change at lower levels of natural law precepts and even more at the level of positive law. (Admittedly, some popes, perhaps especially Pope John XXIII, wholeheartedly praised some of the changes associated with the so-called welfare state.)

CRITIQUE

It is all too easy to criticize natural law theory, whether in itself or as a foundation of a conservative political philosophy. Two fundamental criticisms have already been alluded to. Deontologists claim that the natural law "desire for happiness" is inadequate as a basis for a metaethics that claims to guarantee goodness of intention. Utilitarians say natural law is inflexible, that it makes the good of the greater number suffer in the name of some prohibition of "unnatural" acts. A third objection, perhaps the most popular of all, is that natural law theory is based on an outmoded, unscientific, supposedly universal definition of human nature as always and everywhere the same—indeed, as universally known to be the same in all cultures.

This last criticism might seem to be somewhat blunted by Aquinas' claim, reinforced by John Finnis, that the awareness of basic natural law precepts is not a matter of philosophical expert knowledge but is spontaneously recognized by all reasonably intelligent human beings above a certain level of development. Even so, this puts a rather large burden on the defender of natural law: to prove that, indeed, all humans in all cultures thus instinctively recognize the basic human goods. Some defenders, such as Germain Grisez—one of Finnis's favorite interpreters of natural law—have attempted to do this using data from modern anthropology.[22]

Another popular critique—perhaps less popular now than before—is to accuse natural law theory of committing G. E. Moore's "naturalistic fallacy," i.e., to say that it fallaciously derives "ought" from "is." There is still much discussion in ethical theory circles about the is/ought problem and whether or not the alleged fallacy is in fact fallacious, but in any case it is extremely doubtful that the objection applies to all versions of natural law theory. For example, the Aquinas/Finnis version referred to here does not derive natural law oughts from Aristotelian-Thomistic metaphysical or philosophical-psychological knowledge of Being or of human being. For these two natural law theorists, the basic oughts are *underived*; they are the primitive givens of the moral life, *spontaneously known* by all as natural goods, as the general ends of a moral life.

The principal criticism of natural law theory that I want to make does not address it in its general formulation but as a proposed solution for the social problems of a technological society. That is, my objection is not to natural law theory in the abstract but to the prescriptions for a good technological society of recent popes and Roman Catholic officials—or to the conservative critiques of the technocorporate state of Russell Kirk or E. Digby Baltzell (*assuming* that these two conservative thinkers might ground their critiques in natural law theory).

The first part of the objection is that natural law thinkers are not very imaginative in facing the problems of a technological society—and this in spite of Kirk's glorification of the conservative tradition as rooted in "imagination, not onesided intellectualism."[23] Roman Catholic reaction to the world population problem is obviously the most extreme example. Almost as outrageous are calls for women to "recognize their proper place," or rationalizations of social inequality that too easily—despite protests to the contrary—lend support to racism and other forms of discrimination against the poor. In early modern times, "Jesuitical casuistry" became infamous because certain supposed advocates of natural law were able to find loopholes to justify all sorts of untraditional behavior. But when it comes to the major social problems of contemporary society, such flexibility seems to have deserted natural law—and other conservative thinkers.

The second part of the objection takes the opposite tack. I have suggested elsewhere that solutions to the social problems of a technological culture to which natural law might be relevant would all or nearly all be at lower levels of generality—at the level of so-called "tertiary precepts" or even at the level of positive law.[24] After all, what we are talking about is changed historical and cultural conditions—exactly what Aquinas points out as demanding more specific, *and changing*, applications. At this level it becomes exceedingly difficult to distinguish imaginative approaches following natural law guidelines from any other sort of imaginative approaches. As an example, when conservatives face up to the problem of alienation in contemporary society by appealing to the "principle of subsidiarity" or lauding the "new ethnicism," this is not, as social policy, very different from the "community action" of liberals of earlier decades. In general, at lower levels of generality, dividing lines tend to get fuzzy.

To some conservatives this may seem an unfair criticism, inviting a "damned if we change, damned if we don't" rebuttal. But the criticism is not unfair; it has been invited by every overly abstract thinker since Plato—who opened himself up to an attack of just this sort from Aristotle.

And this brings me to the third part of my objection. Too often, in fact, conservative thinkers, including natural law thinkers, end up as partisans of reaction. Michael Oakeshott, often listed as *the* leading contemporary conservative political philosopher, says that under no circumstances should conservatism become a political creed or partisan platform. Yet it often does—and inevitably of the parties of tradition, privilege, and power. Baltzell is a good example here: though he feels "deference democracy" would be an antidote to the excesses of contemporary antinomianism in an alienated society, in fact it would mean a return—even in his idealized Boston society—to rule by brahmins.

Finally, and perhaps here we come to what is really the heart of the matter, natural law theory and other conservative views are often not, at bottom, "natural" views at all. Natural law is latched onto as a support for what is really a morality based on divine revelation. Aquinas may argue that natural law inclinations are spontaneously and naturally revealed to all, but he admits that divine revelation will aid those with misguided customs or practices to see the light. The popes and other Roman Catholic thinkers are more pragmatic: they appeal indiscriminately to natural law or to the Gospel whenever it suits the occasion. This is equally true of Kirk or Baltzell, for both of whom the church is one of the bastions of tradition, along with family and property.

Having said so much in criticism, and especially having made this last point, I should add one more note. The conservative "voice in defense of spiritual values" should not be discounted or despised in a technological culture. As policy planners, welfare state bureaucrats, and others in the mainstream of an increasingly technological culture go about their business of dealing with contemporary social problems, the temptation is all too real to treat human beings as pawns in a social engineering chess game. In this context, to be reminded of the traditional spiritual dimension of humankind is perhaps *the* key service that conservatives can render to contemporary society. They can also join in protest movements and other activist causes, as some have done on issues of war and peace and economic justice—but in this their natural law principles, despite Murray's claims, seem to offer no better grounding than the grounding other activists have.

15

Liberals: Self-Interested and Progressive

Individualist democracy is rooted in the primacy of the individual human being. . . .

In every important activity—in the arts and sciences, in work and play—it is personal achievement that human beings value most. This respect for private enterprise is applied by individualism to all affairs, including economic affairs. . . .

Individuals work most industriously and happily for themselves, their families and their friends. An economy based on intelligent self-interest is not evil. It encourages thrift and foresight. It avoids the notorious inefficiency and corruption of collectivized economies. It protects open and competitive markets, thus rewarding energetic and efficient producers. It gives effective power to ordinary people by making their choices, as consumers, genuinely sovereign. Private enterprise is the surest path to prosperity, for individuals and for nations.

—Carl Cohen

Socialist democracy is rooted in the nature of the human community. Problems facing an entire community can be solved democratically only when the community acts as a whole. Collective action, therefore, in the economy or elsewhere, is not a departure from democracy but its fulfillment.

—Carl Cohen

I began this book with a quotation from Daniel Bell quoting Friedrich Nietzsche: "For some time now our whole European culture has been moving as toward a catastrophe."[1] Citing Nietzsche as a challenge, Bell gives a nuanced response. He claims to be a liberal, but only in politics; in economics, he favors socialist measures, and in the cultural sphere he says he is a conservative.[2] This suggests the obvious—reinforced in the headnote quotes from Carl Cohen[3]—that one can be a liberal in many different ways.

In this chapter, I take up a final objection to my advocacy of a progressive, activist liberalism as the best response to the challenges of our technological age. The objection comes from within the ranks of liberals themselves, and it is straightforward. Activism as a way of controlling technology, they say, will not work. In saying that solving social problems is a responsibility of professionals as professionals, this approach weighs professionals down with a demand for altruism that we have no right to expect of them.

HISTORICAL NOTES ON LIBERALISM

The liberal label has not been popular recently in the United States, and those who still prefer it are likely to hear themselves described by their enemies as "old-fashioned liberals."

The term "liberal" has had a variety of meanings in the two centuries in which it has been popular. Not long after the Enlightenment, the term became popular in European countries to describe a progressive movement that was generally anti-clerical, anti-traditional, and reformist, but not revolutionary. This was the liberalism, for example, that was opposed by nineteenth-century popes. Benthamites in England can be thought of as defenders of the view, and John Stuart Mill is often taken to be its outstanding spokesperson in Great Britain.

European liberal parties, which had their roots in this movement, tend to favor many different programs, but they can put up a common front, as in the Liberal Manifesto proclaimed at Oxford in 1947:

> We, Liberals of nineteen countries, assembled at Oxford at a time of disorder, poverty, famine, and fear caused by two World Wars; convinced that this condition of the world is largely due to the abandonment of liberal principles; affirm our faith in this Declaration.[4]

The first and third of the manifesto's principles echo Daniel Bell. He says, "I believe in the principle of individual achievement";[5] the manifesto simply affirms the primacy of the individual, saying that the state is no more than an instrument of the community of individuals. These Liberals held subsequent meetings and made later declarations, but they always reaffirmed the same principles, generally coupling them with an invocation of the free market. It is this united front of liberalism that allows Marxists, among other radical critics of the left or right, to attack liberals *en bloc*.

In the United States, the terms "liberal" and "liberalism" have had a checkered history. Only in one area does the liberal label seem to have been applied consistently, to describe a brand of Protestant theology. In the political arena, what later came to be called liberalism originally, around the turn of the twentieth century, bore the label, "progressive." Franklin D. Roosevelt's New Deal coalition of North and South, of ethnic groups and local party organizations of all sorts, of unions and even some business groups (many of these groups not being very reformist by contemporary standards) can be taken as the first professedly liberal movement in twentieth-century American politics.

The euphoria of New Deal liberalism lasted perhaps a generation, through the presidency of Harry S. Truman and beyond. Its rhetoric reached a high point in the two presidential campaigns of Adlai Stevenson, with its highest point in the short presidency of John F. Kennedy. Legislatively, the high point was probably Lyndon Johnson's Great Society program. At no time in this period was this most popular version of twentieth-century U.S. liberalism at odds with cries for "fiscal conservatism," nor were most liberals opposed to the ever-increasing political power of large corporations.

In general, American political liberalism has always been less doctrinaire than the European liberal parties. Nonetheless, by the end of the Johnson years, and especially during the Vietnam War protests and the 1972 presidential candidacy of George McGovern, enemies of liberalism managed to give liberalism a leftist connotation that helped speed the decline of the popularity of the liberal label. By the 1980s, self-styled liberals were often referred to as a "dying breed," stuck, supposedly, with nothing more to offer than remnants of New Deal policies, now nearly fifty years old.

It is not either doctrinaire European liberalism or squabbles about what remains authentic about liberalism in the United States today that concerns me here. What I am interested in—because I want to oppose it—is a view of "the technocorporate state" that is thoroughly centrist. It could be supported by Republican moderates, by Democratic centrists, or by moderates belonging to neither major party. The key point is approval of the widespread influence of large, technology-based corporations on national government.

THE LIBERAL, TECHNOCORPORATE, WELFARE STATE

John Kenneth Galbraith, a critic, characterizes our present system by way of a contrast with earlier stages of our economic and political development:

> In the early stages of corporate capitalism Congressmen and state legislators were the natural servants of the businessman in the affairs of government. The entrepreneur employed resources that were his own. . . . These—money and votes by employees along with the respect and fear inspired by the dependence of the community on his favor or on avoiding his retribution—were thus available to him as political power. . . .
>
> Since World War II, matters have much changed. . . . Services to the planning system have become far more important. The resulting bureaucracies—Defense, AEC, NASA, CIA—are symbiotically related to the planning system. The Executive has come, in consequence, to be an expression of their interest, which is to say the interest of the planning system.[6]

For those who do not know Galbraith's work, "the planning system" (elsewhere called "the technostructure") is his term for the key feature of his post-Keynesian economic analysis. For Galbraith, planning is an essential feature in explaining the economics of the large corporations, with their national and international (as opposed to local) scope of operation and influence.

In the two short paragraphs cited here, Galbraith does not talk about how reform efforts could succeed in controlling the power of multinational corporations as earlier progressive reform movements had succeeded in controlling local businesses in the early stages of the transition to corporate capitalism, but that is the real focus of his book. I will return to those issues later.

It is, then, fairly easy to summarize the main features of a technocorporate state: almost total control of the economy, with only a few pockets of traditional competitive capitalism left among small businessmen and the underground economy; enormous political power of corporate lobbyists and PAC campaign givers closely allied to key bureaucratic heads and Congressional committee chairs and staffs; effective control of an educational system aimed primarily at turning out employees for the corporations, and increasingly for high-technology jobs in high-technology companies or service departments of older industries; a virtual stranglehold on the media; and even a significant impact on family life and leisure-time activities.

In general, most Americans, and probably a majority of the citizenry in most developed countries, accept this welfare state as the way the world is today. The more conservative among them may complain about the rising bureaucracy but, when they obtain power, they do nothing to control it—indeed, they often do much to augment military, intelligence, and police bureaucracies. Nowadays, a very large percentage of this broad middle sector of the citizenry worry about huge budget deficits or savings-and-loan scandals, but they seldom translate these worries into crusades to remove the modern government services that favorably affect their interests as they perceive them.

It is this mainstream view of life in the modern world (with special reference to the United States) that I focus on here. It is much too collectivist for doctrinaire conservatives and too resistant to radical reform to satisfy those on the far left. The mainstream view is thus liberal in standing between these extremes. What I am concerned about, with respect to the welfare state, is its relationship to technology.

THE LIBERAL CONSENSUS AND TECHNOLOGY

One self-described liberal who makes the link to technology in an explicit way is Edward Walter, in *The Immorality of Limiting Growth.*[7]

Philosophical Foundations

Walter summarizes "the philosophical development of English liberalism," his point of departure, as a process of gradually coming to recognize four key ideas: (1) a legal concept of personal freedom (Walter calls these the "procedural" rights of life, liberty, and property); (2) a common-law conception of a social contract that requires respect for all citizens; (3) representative government; and (4) the idea that citizens have the right to rebel, to change their system of government if it no longer serves their interests. Walter attributes the synthesizing of these ideas to John Locke, with the latter's emphasis on the natural right of property and the separation of powers to keep government in check. In Walter's account, the natural-rights base was challenged by later liberals, and property was redefined, following Adam Smith, as *capital* rather than just land holdings. A final note was added, according to Walter, with John Stuart Mill's insistence on the absolute character of individual liberty. Government cannot restrict the liberty of individuals even for their own good (except for the protection of society, as in war).

Walter summarizes: "[The basic] Lockean beliefs are retained by liberals: the priority given to personal liberty, the need to limit government authority (tyranny), and the utility of representative government."[8]

While few people would challenge this summary, Walter's later definition of "essential liberalism" is more controversial:

> The fundamental beliefs about human nature that are vital to liberal philosophy are that (1) people are fundamentally self-interested, (2) self-interest points toward happiness, and (3) people, because of biological and social variability, are temperamentally diverse and consequently seek happiness in different ways. . . .
>
> Two necessary means of achieving self-development are thought to be [4] freedom, so that people can pursue their individual paths, and [5] material well-being.[9]

Two features of Walter's liberal theory stand out. One is the demand for "material well-being" as a condition of the employment of individual liberty. This is the basis of his argument against "no-growth futurism," as he calls his target of attack. If, because of pollution, resource depletion, and the population explosion, there is to be no growth in the future, then some people—the poor—are going to be condemned in perpetuity to a state of material non-well-being that will prevent them from exercising their moral liberties. The other special feature is Walter's emphasis on self-interested individualism.

Individual Self Interest

Each of the points in Walter's analysis needs a more careful philosophical elaboration than he gives. Robert Nozick provides in *Anarchy, State, and Utopia*, a more sophisticated Lockean analysis than Walter's. Nozick's view is well known, and I supply only reminders of the main points. Insofar as Nozick's view affects justice, it is concerned with so-called "material principles"—i.e., with what must be superadded in order to give specific or concrete meaning to "formal" justice defined as "giving to each person what is his or her due." Nozick's is a *desert* view: "We think it relevant to ask whether someone did something so that he *deserved* to be punished, deserved to have a lower share." Nozick also says that his is an "entitlement" or individual rights theory, based on a "historical" principle of distributive justice. He goes on:

> Patterned principles of distributive justice [and almost every suggested principle has been patterned] necessitate redistributive activities. . . . From the point of view of an entitlement theory, redistribution is a serious matter indeed, involving, as it does, the violation of people's rights.[10]

What individuals have a right or are entitled to do, on Nozick's account, is to participate freely in a market system, to engage in the free exchange of goods and services without government interference (except to protect individual rights). Although Nozick speaks of his view as libertarian, it is clear that he sees it as a modern-day version of Lockean liberalism. So Nozick would support Walter's emphasis on self interest.

Liberalism and the Free Market

To return to Walter, it needs to be noted that for him as for Nozick, the free market serves as the ideal. Walter opposes "no-growth futurists," among other reasons, because he is convinced that liberalism demands economic growth, and capitalism supposedly ensures that.

It seems that opponents often summarize arguments better than their defenders, and Bernard Gendron, an opponent of liberalism, provides one of the clearest arguments in favor of growth that I know of. According to Gendron, the liberal argument can be summarized this way. It "says that the most basic social evil is economic scarcity. . . . In effect, the . . . argument tells us that we can solve all our major social problems if we are able to solve our economic problems. Specifically, it tells us that we will probably eliminate war, crime, oppression, and the like once we succeed in eliminating economic scarcity."[11] Gendron does not think that this argument has been as effectively articulated by liberal theorists as by himself. Nevertheless, he thinks there are pieces of the argument to be found in the writings of authors as diverse as R. Buckminster Fuller, the science fiction writer

Arthur C. Clarke, John Maynard Keynes, and post-Keynesian political economist John Kenneth Galbraith (discussed in some detail below).

Gendron may be clear about the matter, and his sources may all support a free-enterprise version of liberalism, but other spokespersons for liberalism are less sure about the connection between liberalism and the free market.

Ronald Dworkin, often taken to be the leading spokesperson for liberalism, attacks the identification of liberalism with capitalism:

> Economic license and intellectual liberty must stand on the same footing only if liberty means license; they are plainly distinguishable, and at some point inconsistent, if liberty means independence.
>
> The Supreme Court confused these two ideas, decades ago, when it decided, temporarily, that if the Constitution protects liberty at all it must protect the liberty of an employer to hire workers on such terms as he wishes. Conservatives confuse these ideas when they use "permissiveness" to describe both sexual independence and political violence and to suggest that these differ only in degree. Radicals confuse these ideas when they identify liberalism with capitalism, and therefore suppose that individual rights are responsible for social injustice. Mill's collected works are not the source of that sort of confusion, but its antidote.[12]

The position that Dworkin is here implicitly espousing he finds not only in John Stuart Mill's *On Liberty* but in Mill's works generally. It is the view "that independence of personality must be distinguished from license and anarchy, and established as a special and distinct condition of a just society." Mill, according to Dworkin, is no defender of the untrammeled free market (*à la* Nozick), and he would not (*à la* Gendron) equate liberalism with capitalism.

Another spokesperson for the liberal viewpoint is John Rawls; he is even sometimes described as the author of "the liberal theory of justice." What is interesting here about Rawls's theory is that he goes even further than Dworkin in de-linking liberalism and capitalism. One critic of Rawls, Marshall Cohen, says that for Rawls it is "a crucial task of moral and political philosophy to make clear the inadequacy of utilitarian concepts, and, more important, to provide a persuasive alternative to them."[13] And opposition to utilitarianism in matters of justice can lead to suspicion of capitalism.

Once again, there is little point in rehearsing here Rawls's well known arguments. But two theoretical points are worth mentioning. The essence of Rawls's rejection of utilitarianism (echoed by many others) is that "a lesser liberty" should never be "accepted for the sake of greater economic and social benefits." To do so, Rawls thinks, would violate the liberty that is the first principle of utilitarianism. In opposition to this, Rawls proposes a non-utilitarian principle of "the common good in the form of the basic equal liberties of the representative citizen."[14]

Rawls's appeal to the common good here, it should be noted, is not based on benevolence or assumed social proclivities of individuals. Indeed, he argues that it cannot be: "We now see why nothing would have been gained by attributing benevolence to the parties in the original position" in the hypothetical social contract.[15] Instead, what Rawls assumes, in deference to the "realistic" situation which gives rise to the need for a theory of justice in the first place, is what he calls "mutual disinterest":

> It is impossible, then, to assume that the parties are simply perfect altruists. They must have some separate interests which may conflict. Justice as fairness makes this assumption, in the form of mutual disinterest, the main motivational condition of the original position [or social contract].[16]

Later Rawls adds: "The reason why the situation remains obscure [in cases of conflict] is that love and benevolence are second-order notions: they seek to further the good of beloved individuals that is already given. If the claims of these goods clash, benevolence is at a loss as to how to proceed."[17]

The situation here, then, is somewhat peculiar. Some well known liberals—notably Dworkin and Rawls—are much more willing than Walter to decouple liberalism from capitalism. Yet Rawls at least supports Walter in basing liberalism on something other than altruism. So there is a range of liberalisms on the issue of economics, but all the versions considered so far tend toward the self-interest end of a self interest to altruism spectrum.

Does Liberalism Require Growth?

What about Walter's final claim, that "essential liberalism" requires economic growth? My task here might appear easy. Is it not the case that "everyone knows" that liberalism favors economic growth?—even economic growth based on improvements in technology? Perhaps. But it is worthwhile returning to Walter's text to see how he makes the argument.

Walter begins his final chapter this way: "In writing this book, I have set two main goals: first, to show that industrial society is not a social relic as no-growth futurists maintain, and, second, that liberalism remains a living political tool by which contemporary moral needs can be met." And the first specific recommendation Walter makes is a call for "greater economic equity . . . so that the socially and politically disadvantaged can improve themselves."[18] The vehicle seems obvious: techno-economic growth.

Walter is convinced that freeing the market from government control, as conservatives suggest, is not the way to bring this about. Unfettered industrialists, he thinks, would simply further their own economic interests and would continue to pollute and use up valuable resources.

Government action is needed, he feels, to direct industrial growth toward moral ends. But Walter does not think that government is the answer either. It is too easily bought off by lobbyists for the big corporations. He demands that "liberals must first convince the public of the need to modify industrial growth" to avoid pollution and "fulfill genuine social needs" such as redistribution of wealth and economic opportunities for the poor. This they can do if they abandon revolutionary political rhetoric and appeal to "politically neutral Americans." (Walter recognizes, at this point, that the problems are not uniquely American; he says all the Western democracies should follow the same course.)

Curiously, in making this recommendation, Walter does not refer back to his earlier formulation of "essential liberalism," or to the assumption that humans are fundamentally self-interested. In my opinion, it would be wildly implausible to base an appeal to the "politically neutral" to redistribute wealth and favor the poor on a principle of self interest. So I turn to an entirely different kind of progressive liberalism that makes it the central core of its program to regulate the corporations in the interest of all citizens.

A COUNTER-PROPOSAL: PROGRESSIVE LIBERALISM

John Kenneth Galbraith is commonly taken to be the leading academic spokesperson in the United States for a liberalism that has as its highest priority the control of technology-based corporations. Galbraith's reform program is clearly spelled out in *Economics and the Public Purpose*, where he says that the needed reforms "follow as ever from the problem."[19]

A Reform Agenda

Galbraith begins with structural reforms.[20] These have three parts:

1. The first reform applies to Congressional elections and involves a presumption against re-election:

> Such a presumption against re-election will greatly enhance the likelihood that legislators will reflect contemporary public attitudes. It will ensure, as a matter of course, the replacement of those who, under present arrangements, are co-opted, willingly or osmotically, by the planning system.

2. The second reform applies to the workings of Congress, especially in terms of committees and the seniority system. Rapid turnover in the membership of committees—another result of presumption against re-election—"would help ensure against . . . subservience . . . to the bureaucracy"—itself largely a tool of "the planning system" (see above).

An "end to the seniority system" and the election of committee chairmen by members "would require that the chairman be responsive . . . to his fellow legislators." As it is, "a 'powerful chairman' is, with rare exceptions, a man who exercises power in the Congress on behalf of one of the more powerful public bureaucracies, the most notable case being the Armed Services."

3. The third of Galbraith's structural reforms is a strong Presidency: "An effective President is one who leads and disciplines the bureaucracy to public goals as distinct from those of the planning system."

This three-part program is clearly in line with the reform proposals of citizen activists like Ralph Nader and John Gardner discussed above in chapter 6.

To his three-part structural reform, Galbraith adds seven measures that are needed to carry out the structural reforms:

4. Measures to *equalize power* within the economic system.

No economic remedy is possible, says Galbraith, that does not enhance the power of the market system or reduce the power of the planning system or both.

5. Measures directly to *equalize competence* within the economic system.

Here Galbraith points to particular functions—housing, transportation, health services, artistic and cultural institutions—that do not lend themselves to control by the private sector, even by private-public partnerships. Galbraith admits explicitly that this involves a degree of socialism.

6. Measures directly to enhance the *equality of return* as between the market and the planning system.

As things stand, the wealth and power to be gained by corporate mergers and similar anti-social manipulations of the stock market far outweigh any hope of gain from competitive small business. Twenty years after Galbraith drafted his proposals, this would include, for instance, regulation of the so-called "junk bond" excesses of the 1980s.

7. Measures to align the purposes of the planning system, as these affect the *environment*, with those of the public.

This is the sort of merger of environmental activism with progressive politics discussed above in chapter 10.

8. Measures to *control public expenditures* to ensure that these serve public purposes.

Again, twenty years later, this would involve correcting such abuses of the Reagan years as the HUD scandals or the savings-and-loan scandal—not to mention the Reagan Administration's massive arms build-ups.

9. Measures to eliminate inflationary tendencies consistent with greater equality in *income distribution*.

Galbraith's long-standing views on these issues are well known.

10. Measures to ensure the inter-industry *coordination* of which the

planning system is incapable.

This again seems to imply a measure of socialism.

Details of this decalogue for citizen activism to control "the technostructure," the corporation/government cooperative arrangement he calls "the planning system," Galbraith then takes seventy-two more pages to spell out fully.

Discussion: Progressive Liberalism and Economic Growth

Galbraith, as noted, says some of the items in his reform agenda will require "a measure of socialism"; and conservative critics routinely accuse Galbraith and like-minded liberals of being socialists. From the opposite direction, radical critics like Bernard Gendron (cited earlier) lump Galbraith together with other growth-oriented liberals as if there were no differences. The truth of the matter seems to be that Galbraith is a defender of the market capitalism of small firms, and that he accuses "the planning economy"—the multinational corporations in collusion with key bureaucrats and Congressional chairpersons—of masking a kind of state socialism under the deceptive slogan of free enterprise when it is not that at all. What Galbraith clearly favors is regulation of the multinationals in a way that would return our political system to some semblance of democratic accountability. To me, this does not represent any simple-minded position for or against the link between liberalism and capitalism. What the position is clear about is that, as an ideal, economics should be subordinate to democratic politics.

What about Walter's claim, discussed at length earlier, that "essential liberalism" requires economic growth in order to assure that the poor have the minimum economic means to exercise their fundamental freedom of choice? In the cited passages, Galbraith is not clear either about such an income floor or about whether it should be provided by government or as a beneficial spillover. (He would, clearly, never call for a "trickle down" of capital-driven growth.) Nevertheless, in other places Galbraith has been clear, as have been many other progressive liberals, that the provision of a decent minimal income allowing the poor to exercise their freedom of choice is a collective concern, not a matter of individual charity on the part of the wealthy. On the other hand, progressive liberals do generally believe that a regulated economy will grow—indeed, that it will grow at least as fast as does the current pseudo-free enterprise public-private partnership. Furthermore, most progressive liberals believe a regulated economy that takes seriously the threats that technological development often represents relative to the environment will *in the long run* be more productive than a pollution-plagued economy. And, finally, most progressive liberals believe that "the military-industrial complex" or defense-contract "welfare for rich corporations" represents a significant drag on economic growth. So here again the picture is not simple, though progressive liberals for the most part seem to agree that a healthy economy, freed of excessive military outlays and sweetheart government contracts for some of the biggest corporations, can afford both environmental safeguards and a subsidized minimal income for the truly needy.

Progressive Liberalism and Self Interest

Galbraith would not be expected to raise philosophical issues, but it seems to me that his collectivist measures with respect to "public goods" such as housing and cultural institutions, along with his strong regulatory stance with respect to limiting the activities of large corporations, are much easier to ground in an ethics of altruism than in an ethics of self interest.

For some years now there has been a lively debate, in analytical philosophy circles, about the relative merits of altruism and self interest in the grounding of ethics. A well known defense of altruism is that of Thomas Nagel in *The Possibility of Altruism*. Nagel is clear about his goal:

> My aim is to discover for prudence and altruism, and other motivations related to them, a basis which depends not on desires, but rather on formal aspects of practical reason.[21]

Nagel's view is based explicitly on a Kantian approach, and his principal opponents are followers of David Hume. This is the way Nagel ends his book:

> The word "possibility" occurs in the title of this book for a reason. Even though altruistic motives depend not on love or on any other interpersonal sentiment, but on a presumably universal recognition of the reality of other persons, altruism is not remotely universal. . . .
>
> To say that altruism and morality are possible in virtue of something basic to human nature is not to say that men are basically good. Men are basically complicated; how good they are depends on whether certain conceptions and ways of thinking have achieved dominance.[22]

This Kantian approach is, of course, debatable, and I will return shortly to a view which challenges it as incapable of providing the guidance we need in solving real-life ethical problems. Meanwhile, I want to cite another attack on self interest by Nicholas Rescher.

Rescher again is clear about his aims—and particularly about the target of his attack:

> The central aim of the book is to expound a case for the significance and

> fundamentality of moral considerations. I want to argue—against a vast host of economists, game theorists, and decision theorists—that one should not take the view that rationality conflicts with morality. . . . And I wish to contend against a vast host of philosophers and social theorists that morality is not simply a product of intelligent selfishness, or self-interested rationality in an unfamiliar guise.[23]

Though I am not especially interested here in the details of this controversy among analytical philosophers, it is probably worth a brief pause to summarize Rescher's argument against what he takes to be the dominant view today. He first summarizes the opposing view in terms that would be appealing to utilitarians as well as economists:

> It would . . . be very convenient if something as inherently problematic and intractable as moral theory could be extracted from something as relatively straightforward as human satisfaction.

This simplification, however, Rescher says will not work:

> Morality cannot rest at the limits of what is, in the final analysis, an individual's private standpoint; it must adopt and implement the social point of view.[24]

Rescher does not end with this bald appeal to the social. He shows himself to be aware of the appeal of self interest, and he continues:

> Internalization of the weal or woe of others . . . presents a way of narrowing the potential gap between morality and interest. Insofar as the welfare-status of others becomes part of our own (in the positive mode), the effect is that our own interest—thus amplified—comes to encompass the moral requisites of promoting the welfare of others.[25]

Rescher then concludes: "In sum, it emerges as a duty . . . to work for a society in which both moral virtues and the virtuous themselves can thrive."[26]

Where Nagel traces his view to Kant, Rescher goes back to Hegel, and neither of them, in all probability, expects his argument to convince that "host" of opponents Rescher lists—from economists to game theorists to sociologists and social psychologists to utilitarian (and other) philosophers. Neither do I mean to say that I think any argument based on altruism can be proven superior, once and for all, to arguments based on self interest. Indeed, this brief discursus on the altruism/self interest controversy has been brought in only tangentially to support my opinion that progressive liberalism is more compatible with altruism than with a self-interest-based morality.

SOME CONCLUSIONS

It is time now to return to my main point. I have been claiming throughout this essay that progressive activism has a better chance than any alternative of dealing effectively with the main problems of our culture that are associated with technology. I have also been maintaining that it is a part of technical professionals' responsibility to get involved in this sort of activism. This is what brings up the question of altruism. Professionals who would limit their duties to narrow technical responsibilities often seem (at least to me) inclined toward a morality linked to self interest—often enough, an interest in saving their individual souls in an afterlife! On the other hand, professionals who see professionalism as involving broader social responsibilities tend to have a broader sense of morality as well.

Throughout this essay, I have linked my faith in this broader understanding of morality and social responsibilities to the approach of George Herbert Mead and John Dewey. Without providing any fully articulated arguments that would satisfy an analytical philosopher,[27] Mead and Dewey consider altruism to be a necessary prerequisite of progressive social problem solving.[28] They agree that self interest is inadequate as a motivation for people to get involved in problem solving, just as it is inadequate as an explanation of the collaborative problem solving that, historically, has characterized progressive social groups. On the other hand, Mead and Dewey also reject deontological or anti-utilitarian ethical theories that base morality on formalism or structuralism (along the lines of someone like Nagel, as cited earlier). Structuralist or formalist demands to do one's duty in creating a more rational society have no force unless they smuggle in some non-formalist content in the guise of ideals or goals or social aspirations. And people do not work to achieve abstract goals or ideals; they fight to see particular values win out over competing values. Further, they do this most progressively in a democratic order that strenuously upholds the rights of minorities, in a society open to all values.[29] Both Mead and Dewey saw privileging the values of some—whether of a particular religious denomination or social class or even a democratic majority—to the exclusion of the values of others as an obstacle to progress. Only an open society can be a genuinely progressive society.

It is this sort of belief in an ever-expanding democracy—a democracy that strives to include an ever-wider circle of values—that seems also to motivate progressive activists, including activist engineers, scientists, physicians, and other technical professionals. At least this is the sort of motivation I find most often in the representatives of the technical professions I see getting involved in activist causes. Self interest seems the most *unlikely* explanation of their behavior.

We live in troubled times. Almost everyone agrees with that. We are faced with enormous problems in every sphere of contemporary life—the family, health, education, politics, the environment, and so on and on. As a democratic society, we have made a compact with the professions. In return for certain specified privileges, we expect help from them in solving urgent social problems. As a result, it seems to me to make sense that we should expect of *every* professional a *primary* commitment to the solving of social problems. I do not deny the good will of those who say they are doing their part to serve the public and protect it from harm by doing their technical best. I only question their understanding of what "technical best" means. I believe society expects more of professionals than mere technical competence. In any case, the *best* societies seem to be those that can depend on their social problem solvers to feel the pull of *social* responsibilities.

I want now to end this chapter and the main part of my book with a sort of decalogue of such social responsibilities. In deference to the self-interest liberals as having the strongest of the objections to my approach—altruistic activism will not get the job of challenging technology done but self interest will—I am going to provide two formulations of each obligation. One, obviously, will be based on beneficence or benevolence. The other will be based not on self interest but on something like Rawlsian justice-as-fairness, with its assumption (see above) of "mutual disinterest," indeed of conflicting interests.[30]

A note. It may be that the benevolence version of each principle will seem better suited to characterize the social responsibilities of individual technical professionals, and that the justice (compensation) version better suits the institutions with which, in our technological society, professionals are so often associated—professional societies, technology-based corporations, technical bureaus in government, etc.

Each of the principles or calls to reform that follows can be viewed as an instantiation of a single basic principle or a global call to activism. The basic principle is simple and straightforward: *If you (group or individual) have contributed to the creation or the continuation of a problem, you have at least some responsibility to help solve the problem.* A Mead-Dewey type of formulation—one that appeals to benevolence rather than to obligations or principles—might go something like this: *There are great social evils in our society; together we can solve them; so we should get started.*

1a. The biotechnology revolution has not generated the great risks to the public that alarmists warned about, but it has generated a number of specific problems that are a cause of great concern: regulatory confusion, conflicts of interest (or at least the perception thereof), reinforcement of the injustices of large agribusiness enterprises relative to small farmers, or of multinational corporations' exploitation of third-world peoples. (See chapter 7, above.) Therefore *biotechnology researchers, entrepreneurs, and their organizations have a responsibility to deal with the problems they have helped to create.*

1b. The biotechnology revolution is often touted by scientists as proof that scientists are motivated by a desire to protect the public, so the corresponding benevolence call to action here would be: *Biotechnology represents a golden opportunity for scientists, bioengineers, biomedical researchers, etc., and their organizations to demonstrate to the public the high level of their commitment to the public good rather than to self interest.*

2a. The "electronic revolution" is a slogan often invoked to promote the positive goods that computers, automation, expert systems, and so on, can offer to the society of tomorrow. Meanwhile, electronic devices have made possible the greatest threats to civil liberties since the drafting of the Bill of Rights. (See chapter 8.) Therefore *computer and electronic professionals, and their institutions, have a responsibility to examine the civil-liberties implications of every development in their field and to consider civil-liberties protection to be as important as technical development.*

2b. The traditional groups that take it as their primary concern to educate the public and *to safeguard civil liberties*, in the courts and elsewhere, have every right to expect that *well-motivated scientists and engineers will work with them* and not against them. On many occasions, this may entail activism against organizations—even governments—that are misusing electronic devices.

3a. The greatest threat of all to our society remains that of nuclear war, even after the end of the Cold War. Furthermore, the greatest pollution threat of all comes from the nuclear industry, either in terms of Chernobyl-type releases of radiation or by a failure to solve the problem of permanent disposal of high-level nuclear wastes. Therefore *the nuclear industry, public and private, and every worker in it, has a very great responsibility to protect the public from the harms of radiation.*

3b. Many anti-nuclear activists perceive the nuclear industry's current efforts to fulfill this obligation to be totally inadequate. This represents a most interesting area to appeal to the *benevolence of nuclear professionals* to work collaboratively toward a reconciliation between the fears of the activists and their honest efforts to protect the public. Collaboration is much more likely to work, in the long run, than confrontation. (These issues are addressed in chapter 9.)

4a. Lesser threats to the environment than nuclear irradiation abound. Some people even worry about threats to the entire ecosystem or to life on earth. Since most of these threats arise from technological development, *the corporations and governments, and their employees, responsible for pollution have an obligation to stop polluting and to clear up the damage already done.*

4b. Environmental activists, like anti-nuclear activists, are generally perceived by industrial spokespersons as the enemy. *Benevolence on the part of governmental and industrial proponents of technological development* would again suggest a collaborative rather than a confrontational approach. The ecosystem nourishes all of us. (These issues are addressed in chapter 10.)

5a. As noted in chapter 4, above, a number of engineers and engineering educators, along with others, have recently mounted a crusade for a technological literacy that will go beyond traditional calls for scientific literacy and help prepare citizens of the future to be competent decision makers on technological issues confronting society. A compensatory justice obligation under this heading is controversial, depending as it does on the claim that the corporate culture has been, for decades, heavily involved in setting school curricula. So a compensation obligation should read like this: *To the extent that technical professionals and their organizations have failed to see that technological literacy was included as part of the school curriculum in the past, they have an obligation to see that it is included now.* Defenders of this obligation would often add a second: that the United States (and presumably other technologically-developed countries) has an obligation to generate enough technologically literate citizens for us to remain technologically competitive. (The more extreme would say to retain our lead in technology.)

5b. A benevolence need here is much broader than the call for technological literacy. The most enlightened reformers, on the education issue, seem to be those calling for the education of *well-rounded citizens who can put their new technological literacy in a broader context*. Collaboration between these interdisciplinary humanities reformers and technological literacy reformers is a noble goal in the disastrous contemporary education scene.

6a. If education is in crisis, health care is worse—or at least that seems the near universal consensus. Furthermore, many reflective critics, including medical school educators and students of medical education, recognize technology as an important negative factor relative to compassionate, comprehensive, and equitable provision of health care. (See chapter 5.) To that extent it would seem to be *an obligation incumbent on medical educators and the medical establishment generally to redress the imbalance between high-technology specialization and urgent primary care, especially for the neediest members of society.*

6b. Many spokespersons for the medical profession maintain that *benevolence*—the good of the patient—has always been *the hallmark of the physician*. What that requires, under this heading, would be a call for young physicians to recognize the priority of benevolence over the money to be made in high-tech specialties. At present this does not seem to be the case, though surely idealistic young would-be physicians ought still to respond to the call of benevolence, if it can be effectively articulated.

7a. Another area of widespread concern in our society has to do with the world of work: mind-numbing meaninglessness on the job and the perpetual threat of loss of jobs as companies close plants in one location and move elsewhere. Some critics (see chapter 13) blame much of the problem here on automation; conversely, industrial spokespersons make excuses for their companies based on the need for automation "to remain competitive." In either case, there seems a *clear obligation in justice to compensate workers who are victims of "technological progress."*

7b. A benevolence perspective, on this matter of meaningful work and plant relocations, would appeal to the *commitment (of the best companies) to workers and their communities*. This is the high ideal held up by Edmund Byrne, as reported in chapter 13—though Byrne is realistic enough to believe that the ideal will only be realized by collective efforts of unions and their community supporters, perhaps using the law, appropriately changed, to enforce the obligations in 7a.

8a. Has democratic politics been corrupted by the media—or, something more sinister, by corporations controlling the media? There is probably very little consensus on this matter, though there is widespread concern over current abuses of the democratic process. So I will formulate this principle hypothetically: *If it can be established that corporations are deliberately controlling the media in order to control the electorate, citizens have an obligation to resist this control as vigorously as they can.* (Similarly, professionals in the media or in politics would have an obligation not to be part of this anti-democratic conspiracy.)

8b. A benevolence view, here, would probably be less cynical, focusing on the *noble calling of citizen involvement in public affairs*—perhaps stressing the need to educate ever broader segments of the public to get involved, including education about the power that an alert citizenry can wield. With regard to the hypothesized cynical control of the electorate through manipulation of the media, this would just be one more area for involved citizen activism. (These issues are raised in chapter 6.)

9a. In talking with others about my project in this book, and letting friends read drafts, I have run into the objection that chapter 3 does not belong—that problems of the family are not peculiar to a technological culture. So I will formulate this ninth principle, like the eighth, in a conditional way: *To the extent that family life is made more difficult by technological development, to that extent corporations, government agencies, and their employees have a responsibility to become involved in the solution of family problems.* This, presumably, is the sort of obligation appealed to in calls for family leave policies, for alcohol and drug counselling as a part of employee relations, and similar measures.

9b. A benevolence perspective would simply shift the emphasis here, appealing to social obligation motives rather than to enlightened self interest. (The need for activism in this effort is underscored in chapter 3.)

10a. I began this book with references to "the crisis of technological culture." Under that heading, conservatives lament the decline of standards, and radicals prophesy the end of culture, not only a decline but a fall into total nihilism. Since claims about the end of the world, or the decline of the West, or the end of (science-and-technology-based) modernity, are such vague and sweeping generalizations, it is hard to formulate a technology-related compensatory-justice principle or obligation in connection with them. The best-known formulation of this sort, Hans Jonas's famous "ethics of fear"—in matters technological, "Moral philosophy must consult our fears prior to our wishes to learn what we really cherish"[31]—is presented as a sort of neo-Kantian categorical imperative for the modern world threatened by nuclear destruction, ecological collapse, or bioengineering threats to human nature.

10b. It is as a response to this sort of technological doomsaying that I have written this book. All the best recent interpreters of John Dewey and George Herbert Mead tell us that these two great thinkers would not have succumbed to fear in the face of technological threats to our culture.[32] Indeed, Cornel West says Dewey, were he alive today, would view our age not as the end of the modern world but as a unique opportunity for multicultural development—and this not only in opening up the universities but in providing the opportunity for people to do social problem solving rather than *fin-de-siècle* theorizing.[33] But this will not happen, and people will not struggle to make ours a more inclusive culture, unless we make our appeal on the basis of altruism and social responsibility.

Perhaps scientists and engineers and biomedical researchers and all the rest of the technical community will respond less quickly to this vague call to cultural reform than to any of the other claims of obligation or calls to benevolence that I have listed here. But perhaps also that is to be expected. Meeting the challenge of a technological world calls for action on many fronts. In my view, even the New Progressive Movement occasionally referred to here would still have to attack problems piecemeal, one at a time. Contrary to those totalistic thinkers who call for radical reform, I believe the best response to the challenge of technology is to motivate as many activists as possible—including technical professionals—to deal with local manifestations of the ills of our technological world one at a time, or if several collectively, then one after another. Reform, not revolution, is our best hope.

16

Academic Philosophers

A problem of applied ethics . . . has become especially acute with the technological developments of recent decades: Do people have a moral and political right to privacy? Philosophers [with the special qualifications which analytic philosophers can bring] have recently contributed to public discussion of this and many other practical questions by exploring the conceptual issues and doubts which sometimes make policy choices more difficult.

—Thomas D. Perry

It has been stated [here] that philosophy grows out of, and in intention is connected with, human affairs. . . . [This] means more than that philosophy ought in the future to be connected with the crises and the tensions in the conduct of human affairs. For it is held [here] that in effect, if not in profession, the great systems of Western philosophy all have been thus motivated and occupied.

—John Dewey

Vital and courageous democratic liberalism is the one force that can surely avoid . . . a disastrous narrowing of the issue[s]. . . . The question[s] cannot be answered by argument. Experimental method means experiment, and the question[s] can be answered only by trying, by organized effort. The reasons for making the trial are not abstract or recondite. They are found in the confusion, uncertainty and conflict that mark the modern world.

—John Dewey

In chapter 15, I was satisfied to say that, as a motivator for the sort of progressive activism I am championing in this book, altruism is more plausible than self interest. In making that point, I referred—however briefly and inadequately—to the body of literature by academic philosophers on altruism. Many academics would say that if I want to defend altruism as a basis of the view I espouse, it is incumbent upon me to make the argument in an academically respectable way. I do not wish to do that, and in this chapter I attempt to explain why.

My claim throughout the book—following John Dewey and George Herbert Mead as I understand them—has been that progressive democratic social activism is the best available response to the challenge of technology. Many of my colleagues in academia will wonder about that. What, they might well ask, about *applied ethics* approaches to the solution of particular technosocial problems? Activism seems opposed to the true spirit of philosophy—at least that is what George Allan seems to be saying in *The Realizations of the Future: An Inquiry into the Authority of Praxis*[1]—and to set activism against academic applied ethics seems almost perverse at a time when so many applied ethicists are addressing the same issues I address in this book.

I begin the chapter with reference to a very strong claim that analytical philosophy has unique qualifications when it comes to dealing with urgent contemporary issues.[2]

THE CASE FOR APPLIED ANALYTICAL PHILOSOPHY

It has been proposed by Thomas Perry that analytic philosophers have "special qualifications" that they alone can "bring to the clarification of public issues."[3] Perry is clear about what his ideal is: "Philosophers are not content to leave the great metaphysical questions to poets, dramatists, and novelists, who examine them *primarily* for the purpose of art rather than for the pursuit of literal truth."[4] He also has his heroes: Bertrand Russell, W. V. Quine, Keith Donnellan, and Saul Kripke—who, he thinks, have produced the purest attempts to utilize the latest findings of science together with logical techniques to arrive at the literal truth.

Perry counts others as "serious" philosophers even though they recognize that some traditional philosophical problem areas—in ethics, aesthetics, political philosophy, and philosophy of law—can only be dealt with in terms of "elucidations," not "exact analyses." Here he lists more heroes: R. M. Hare, Kurt Baier, William Frankena, and John Rawls.

Perry even admits that "there is more than one way to be serious in philosophy," where he sets up a contrast with philosophers "in the speculative tradition . . . [of] continental Europe." About these philosophers he says nothing on the score of whether they can contribute to the clarification of public issues.

Perry's precise examples of "elucidations"—which, because they build on one another with the quality of scientific discourse, therefore contribute to the clarification of public issues—are not actually drawn from his heroes, either the Russellians or those espousing "the moral point of view," but from younger followers in the latter tradition. "Do people have a moral and political right to privacy? Philosophers have recently contributed to public discussion of this and many other practical questions by exploring the conceptual issues and doubts which sometime make policy choices more difficult." What Perry deals with under this heading is a series of arguments by Judith Thomson, Thomas Scanlon, James Rachels, and Jeffrey Reiman which, he says, "throw increasing light on the privacy problem."[5]

Though Perry seems to think his goal requires no definition, one of his heroes, Quine, provides a clear explication that his predecessor Russell would have approved. Quine says that particle physics might offer the clearest example of literal truth: "Two sentences agree in objective information, and so express the same proposition, when every cosmic distribution of particles that would make either sentence true would make the other true as well." But of course Quine adds that we "can never hope to arrive" at this ideal, so empiricists have substituted for particles the "introspection of sense data," or if they are "more naturalistically inclined, . . . neural stimulation."[6] And, again as everyone knows, Quine does not think even these latter, lesser goals can be achieved.

It is plausible to cite Russell as the source of the ideal of analytical philosophy—an ideal that quickly got transformed into an ideal for *academic* analytical philosophy. In a popularized statement of his ideal, Russell claims that analytical philosophy, with its reliance on logical techniques, is "able, in regard to certain problems, to achieve definite answers." In this, he says, its methods "resemble those of science."[7]

Hans Reichenbach has provided what is perhaps the clearest statement of another aspect of this ideal—its close connection to the empirical sciences. Reichenbach says it is up to "the old-style philosopher to invent philosophical systems, for which there still may be a place assignable in the philosophical museum called the history of philosophy"; in place of this, Reichenbach says the "scientific philosopher" goes to work at serious analysis of the findings of science.[8]

In this chapter, I explore two objections (I hope they are "serious") to this proposal. The first objection to Perry's claim about the unique qualifications of analytical philosophy has to do with the boldest part of the claim—that applied analytical *ethics* can make unique contributions to the clarification of pressing public issues.

ANTI-ACADEMICISM

The view that academic applied ethics has unique qualifications would be opposed by many anti-academic philosophers. In *One-Dimensional Man*, Herbert Marcuse accuses analytical philosophers of being reactionary guardians of the *status quo*.[9] In *The Technological Society*, Jacques Ellul views analytical approaches of all sorts as "technicism" or academicism with absolutely no potential for dealing with the urgent public issues of our technological world.[10] And, of course, in *Philosophy and the Mirror of Nature* and *The Consequences of Pragmatism*,[11] Richard Rorty, once himself a leading analyst, has declared that philosophy in the analysts' sense (and perhaps in any sense) is dead.

As everyone should know, it was the view of John Dewey—most notably in *Reconstruction in Philosophy*[12] and *Liberalism and Social Action*[13]—that philosophy grows out of and *must* contribute to the solution of urgent social problems. If a particular approach to philosophy does not contribute to the solution of social problems, Dewey brands it as academic, in a pejorative sense, or blames particular philosophers for contributing to the problems that block progress by clinging to outworn dogmas or persisting in the "quest for certainty." Dewey would invite all philosophers to join with other intellectuals in organizing to employ the "experimental method" to deal with "the confusion, uncertainty and conflict that mark the modern world."

George Herbert Mead certainly shared this view with Dewey, but in at least one place he gave it explicit expression in a way that Dewey did not. According to Mead, the progressive creative problem solving of a community is, by the very fact of its being progressive, an *ethical* endeavor. (Indeed, in his early delineation of the various philosophical disciplines as parts of the social *praxis* of a progressive community, Mead defines ethics in just that way.)[14] In a passage I have quoted repeatedly in this book, Mead says:

> The order of the universe that we live in *is* the moral order. It has become the moral order by becoming the self-conscious method of the members of a human society. . . . The world that comes to us from the past possesses and controls us. We possess and control the world that we discover and invent. And this is the world of the moral order.[15]

For both Dewey and Mead—and, before them, William James[16]—this necessarily-ethical progressive social problem solving involves the sort of openness to and tolerance of *all* points of view in a democratic society which is usually associated today with civil-liberties liberals of a progressive sort. (I will return to this point in my conclusion.)

For James and Mead and Dewey, it was enough to define a social movement as progressive if it aimed at the improvement of society by democratic means. Today, we have become more skeptical, perhaps even cynical, about progress.[17] Still, I think the Pragmatists' view of social movements as progressive if they help to remove particular "problems that block progress" in the broad historical sense is clear enough, if vague. And perhaps their optimism is preferable to our current mood of cynicism.

That philosophers ought to contribute to this progress was Dewey's main point in *Reconstruction in Philosophy* and *Liberalism and Social Action*. Like all intellectuals facing the urgent crises of the day, we philosophers have our contribution to make—and that includes philosophers of all persuasions, not just analytical philosophers. The only condition is that we seek to contribute to progress, to the solution of *real* problems that vex not only our own society but of humankind world-wide.

OBJECTIONS TO POSITIVISM

The second objection to Perry is related to the first, though the connection requires a little elaboration. It seems that, for Perry, epistemology is in some sense "purer" than and the basis of analytical or theoretical ethics—which, in turn, is the basis for applied ethics. Both Dewey and Mead, by contrast, hated epistemology—almost *any* version of epistemology, but especially positivism.

Rather than emphasize the epistemology of science (Reichenbach's ideal), Mead would—along with Thomas Kuhn,[18] but long before him—focus on particular scientific communities, where he makes the "social act" (i.e., the *praxis*) of a community the object of analysis. According to Mead, "the unit of existence is the act."[19]

Mead illustrates this in his attack on all epistemological accounts of science and its history in "Scientific Method and Individual Thinker."[20] Mead attacks Kantian transcendentalism, Hegelian idealism, rationalism, neo-realism, positivism, and the empiricism of Hume, Mill, and Russell. In most cases he is kind enough to recognize the positive contributions of each epistemological theory, but he maintains that all fail to grasp the primary importance of the social act, the *praxis* of the scientific community at any particular point in history. For instance, attacking Russell, Mead says:

> It would be impossible to make anything in terms of . . . sense-data and of symbolic logic out of any scientific discovery. Research defines its problem by isolating certain facts which appear for the time being not as the sense-data of a solipsistic mind, but as experiences of an individual in a highly organized society.[21]

Again, opposing positivism and something close to behaviorism (or what later came to be called "methodological individualism") Mead says:

> For science . . . particular experiences arise within a world which is in its logical structure organized and universal. They arise only through the conflict of the individual's experience with such an accepted structure. For science individual experience *presupposes* the organized structure; hence it cannot provide the material out of which the structure is built up. This is the error of both the positivist and of the psychological philosopher, if scientific procedure gives us in any sense a picture of the situation.[22]

In short, whatever various epistemologies of science can offer, it is never the total picture; they fail to situate their insight within the social act of science as it has been historically practiced—that is, within the *praxis* of an organized group of scientists whose world-taken-for-granted is shaken up by some anomalous and problematic discovery which then sets the community on its way to establish another world, and so on and on.

Recent opponents of positivism have taken a somewhat less abstract approach. I will summarize four, beginning with Kuhn (who might be said to be most similar to Mead).

a. It has become a cliché to date the recent historicist/anti-positivist movement in philosophy of science from the publication of Kuhn's *The Structure of Scientific Revolutions* in 1962, so I will bow to custom and also begin there. By now, thirty years later, the main thrust of Kuhn's attack on the positivism of both Carnapian inductivism and Popperian falsificationism is well known. Kuhn's conclusion—that he is describing a "process that should somehow, in a theory of scientific inquiry, replace the confirmation or falsification procedures made familiar by our usual image of science"[23]—is based on his reading of major revolutions in the history of the physical sciences from the Copernican revolution in astronomy to the Einsteinian revolution in physics. The trigger of a Kuhnian revolution in pure science is what he calls a crisis: some anomaly, usually of a theoretical sort, looms so large in the scientific community that it can no longer be ignored. This seems far removed from the applied science I have focused on in this book, and Kuhn's sort of intellectual crisis would not seem to be what motivates concerns today about the negative effects of science and technology. But there is an aspect of Kuhn's historicist approach that I want to dwell on briefly. What I have in mind is Kuhn's insistence on learning how to do scientific work by apprenticeship in a particular scientific community. Kuhn says: "If this book were being rewritten [rather than just adding a postscript], it would . . . open with a discussion of the community structure of science, a topic . . . [of recent] sociological research . . . that historians of science are also beginning to take seriously."[24] In these communities—and Kuhn says they typically number under a hundred members, often far fewer—students learn by doing problems, and in doing them they are "learning the language of a theory." At the same time, they are "acquiring the knowledge of nature embedded in that language . . . by a process like ostension."[25] There is no way, Kuhn feels, that this could be explained in positivist terms.

b. Paul Feyerabend's "epistemological anarchism" as espoused in *Against Method*[26]—his claim that, in science, "anything goes"—is also obviously relevant. To those familiar with what goes on in applied science, when the push is on to beat the competition—whether an enemy in wartime or just beating a competitor to market—"anything goes" might seem an ideal explanatory device. On the other hand, Feyerabend's primary intent seems certainly to have been to cast doubt on the purity of so-called pure science. If one were to take seriously this anarchist mode of (non?) explanation of scientific progress, the upshot might be a reinterpretation of academic science along the more helter-skelter lines of industrial and other "targeted" or "mission-oriented" research.

c. Among recent post-positivists, John Ziman has proposed a

sociological "radical epistemological relativism" that applies to all knowledge but has special implications for academic science. Ziman lays special emphasis on "big science and technology." According to him,

> "Public" knowledge studied in archives such as libraries . . . has a major influence on the "private" knowledge within individual minds. A consensual view is usually introjected into the mental worlds of most scientists in a particular generation, producing the resistance to paradigm change pointed out by Fleck and Kuhn.[27]

But Ziman's main emphasis is on real-world science:

> Science is valued by ordinary citizens, by powerful individuals such as politicians and company directors, and by corporate bodies such as commercial firms and government agencies, primarily for its *use*. It is fostered mainly as a resource to be applied to the furtherance of individual and/or collective activities whose goals are *not specifically the advancement of knowledge*.[28]

This Ziman contrasts with academic science, saying that the popular understanding

> . . . takes no account of several attributes of science that often motivate scientists personally, such as . . . the aesthetic satisfaction of "discovering and explaining the marvels and mysteries of the world about us."[29]

Both accounts are products of Ziman's radical *social* epistemology of science and technology, and it seems thoroughly inconsistent with positivism.

d. If Kuhn's and Feyerabend's and Ziman's claims were not enough to exasperate orthodox positivist philosophers of science, a new breed of sociologists of science have turned out to be even more nettlesome. Here is Rom Harré's summary of the new "anthropological" approach to the study of science:

> Laboratories are looked upon with the innocent eye of the traveler in exotic lands, and the societies found in these places are observed with the objective yet compassionate eye of the visitor from a quite other cultural milieu. There are many surprises that await us if we enter a laboratory and study a group of scientists in this frame of mind. The idea that the enterprise can be defined in terms of an idealized epistemology, whether that of experimentally based inductions or of the conjectures and empirical refutations of the logicist [philosophers of science], is quickly refuted.[30]

Karin Knorr-Cetina and others, such as Bruno Latour and Steve Woolgar,[31] are saying that, in order to examine science and technology in the most objective light possible, we must take up Kuhn's suggestion that scientific communities ought to be the unit of analysis. They then push this suggestion about as far as it will go, with a resultant view of science that—in addition to what Harré says about the view being anti-confirmationist and anti-falsificationist—is radically relativist, relativized to particular communities at particular historical junctures.

THE QUESTION OF RELATIVISM

If we reject old-fashioned positivism of the Logical Positivist or Logical Empiricist sort, must we go all the way to the opposite extreme of relativism? Joseph Margolis does not flinch from such implications. In *Pragmatism without Foundations: Reconciling Realism with Relativism*,[32] Margolis mounts a massive attack against all forms of foundationalism and in favor of what he calls a "restricted" but also "powerful" form of relativism.

Margolis, in denying that his view falls among a list of eight untenable but popular construals of relativism, lays down a set of six conditions that he thinks his version can meet (and that any form of relativism that could be viable today must meet). A viable relativism, he says, must be:

(i) internally coherent;

(ii) intended to account for assigning truth values or otherwise justifying in cognitive terms well-entrenched practices and forms of behavior;

(iii) at least moderately congruent with the tradition of such practices and behavior;

(iv) responsive to well-known and otherwise reasonable conceptual challenges;

(v) set within the framework of an articulated philosophical strategy; and

(vi) dialectically pitted against its own opponents.[33]

With this set of conditions in mind, Margolis defends an anti-foundationalist relativism not only in the obvious fields of aesthetics and morality but in every field of human intellectual endeavor. Margolis begins his defense of this view with aesthetics, but he believes that the relativistic implications of popular accounts of aesthetics (his opponent here is Monroe Beardsley) can easily be extended to other fields. Margolis maintains that, "Beardsley's argument, intended to obviate and even to defeat relativism, actually turns out to provide the solid foundation for a pertinent form of robust relativism."[34] The crux of the matter has to do with being able or not to determine an author's original intent—something Beardsley had earlier said we cannot do (and Margolis references more recent authors who echo this view). From the demonstrability and historical inevitability of the inability to determine an author's intent in literary texts, Margolis goes on to other sorts of texts: "It is not difficult to see that counterpart arguments can be easily mounted regarding the interpretation of language and history and human conduct" more generally.[35]

This opens the way for Margolis's most radical claim: "If there is no convincing way in which to provide a theory of knowledge and inquiry in which inquiry itself is completely transparent, . . . then, globally, there is no way to demarcate the realist and idealist elements of human knowledge."[36] And Margolis draws the obvious conclusion: "We should then have to concede a hermeneutic dimension to all human science, including the physical sciences."[37] A large portion of Margolis's book is then devoted to drawing these relativistic conclusions from the foundations (pun intended) of philosophical conclusions of authors such as W. V. Quine and Thomas Kuhn—however much these authors have resisted the claim that their views have relativistic implications.

In an intriguing chapter, "A Sense of *Rapprochement* between Analytic and Continental European Philosophy," Margolis draws out certain corollaries of his "robust" relativism that make his work relevant to my theses in this book. He begins the chapter discussing Quine, and ends with this paradox (that would be resisted by so many other followers of Quine): "The Quinean program is as much an extravagance as the Heideggerian—and for the same reason: it betrays its own most forceful insight."[38] What Quine's most forceful insight is taken by Margolis to be is its legitimation of pragmatism, of *praxis*.

Much of this chapter focuses not on Quine but on a unique overlap in the otherwise opposed views of Heidegger and Marx. Margolis, despite his anti-foundationalism, retains in his theory a measure of transcendentalism—enough to assure "inquiry a measure of objectivity relativized to the conditions of *praxis*."[39] (A page earlier, Margolis had said relativized to "the stablest technological features of social *praxis*.") And here is the overlap he finds:

> That transcendental arguments . . . are a species of empirical argument . . . is, broadly speaking, the consequence of Heidegger's thesis of historicity; and that our best clue about the validity of such arguments lies with the stablest technological features of social *praxis* . . . is, broadly speaking, the consequence of Marx's thesis about the relation of production and consciousness. The technological, therefore, performs a double role.[40]

The first role Margolis assigns to "the technological" (following Heidegger's and Marx's accounts) is to show how reality must disclose itself—namely, "through social production, invention, experiment, intentional action, and attention to the conditions of survival."[41] The second role Margolis sees for a focus on the technological is to undermine the extravagances Heidegger and Marx add on to their basic insight (Heidegger's pessimism about our technological society and Marx's unhistorical belief in laws of history)—but also the extravagances of the hope that Quine's program can avoid a relativism of the sort Margolis wants to defend. What Margolis's relativism requires is that we recognize that the interpretation of "the physical sciences [can] not afford to ignore the systematic role of the actual historical work of particular human investigators"[42]—and that, therefore, "the human studies" are needed to interpret science and cannot themselves be reduced to a scientific (let alone a Carnapian "physicalist") model.

This digression on the possible relativist implications of recent anti-foundationalism in philosophy (whether analytical or Continental) should not necessarily be read as an endorsement of Margolis's view. (I believe he does make his case persuasively, and so far I have seen no effective rebuttals by his opponents, including such luminaries as—to cite two extremes—Hilary Putnam and Richard Rorty.) The reason I include the digression is to emphasize that at least some philosophers agree that contemporary analytical philosophy is faced with serious internal challenges. It can hardly be taken, as Perry would want it to be taken, as offering uniquely authoritative guidance when it comes to applications of philosophy to contemporary issues.

This, by itself, would not legitimate a Mead/Dewey approach to philosophy as social activism. However, though Margolis does not make the point explicitly, there is a certain compatibility between Margolis's emphasis on social *praxis* and that of Mead and Dewey (especially in Mead's anti-epistemology argument, above).

A CHALLENGE TO RELATIVISM

It is safe to predict, almost with certainty, that academics will step forward to challenge Margolis's relativism. I will not attempt that here. Furthermore, I doubt that Mead and Dewey would disagree in any fundamental way with the thrust of Margolis's anti-foundationalism. (Someone might even say Margolis's book can serve the same purposes, for our day, as Dewey's *The Quest for Certainty*[43] did in his day.) But I do want to express one demurrer.

In chapter 1, I said that I think Cornel West's *The American Evasion of Philosophy: A Genealogy of Pragmatism*[44] admirably captures the activist philosophical thrust of Dewey's pragmatism. And I suspect that West might worry that—in the effort to enlist academics in a Dewey-like crusade to make the universities more multicultural and the world more democratically open to the legitimate claims of "the wretched of the earth," of people heretofore left out of our cultural mix—Margolis's relativism might seem to be a stumbling block.

But there is a more particular way to put the point. Margolis keys his *praxis* orientation, his necessary limit on the possibility of foundationalism, to the survival of the human race at any point in history:

> The praxical does *not* signify a subterranean structure fixed either for a totalized world history or for allegedly distinct but bounded phases of an

> open-ended human history: it signifies only the refusal to disengage science and philosophy from the tacit, biologically grounded impulses of human societies to survive and reproduce themselves.[45]

Or again, more briefly: "It is through social production, invention, experiment, intentional action, and attention to the conditions of survival . . . that our sense of being in touch with reality is vindicated."[46]

I believe West would say, however (along with other recent interpreters of Dewey[47])—and I would certainly say—that our focus should not be simple survival (even of the human race as a whole) but *flourishing*. And in particular we should worry about an ever-greater expansion of democracy, opening up possibilities for an ever-larger number of previously-excluded groups to participate in the search for meaning in our technological world. In simplest terms, a Deweyan progressive pragmatism should be keyed to *justice*, not mere survival. Though Margolis occasionally refers to Dewey in mounting his attack on foundationalisms of all kinds, I am not aware of any of his writings in which he follows Dewey and West in saying that the primary function of philosophy is and ought always to have been the improvement of the human condition (interpreted in terms as democratic as possible).

West's version of an updated and more politically active American Pragmatism is also useful in making another important point. That is, that we must deal with such activist demands of philosophy in our particular cultural context *as determined by historical forces*.

OUR HISTORICAL MOMENT

What is the historical moment that faces us today and ought to involve all intellectuals, including academic philosophers? Even West at his most extreme would not say we have reached a historical point of de-privileging the ruling class with its ideological defenders. But we seem, in our so-called postmodern times, to have reached a level of de-privileging *intellectuals* that has not been seen since the Cartesian era. There is even a degree of willingness on the part of a great many intellectuals to de-privilege themselves, to tone down their claims of expertise, to work, along with others not so well situated, toward the solution of urgent social problems that affect all classes in society from the "first" world to the third. If West would not say so, I think Dewey would, and I certainly would say that we have, today, a unique historical opportunity to *do* something about injustice, about the exclusion of voices from our cultural search for meaning in a technological world.

CONCLUSION

If my view is even slightly plausible, it does not follow that epistemologically-oriented analytical ethicists, theoretical or applied, have nothing to contribute to such activism. But their possibilities for contributing are certainly not unique. All philosophers, of whatever school—indeed, all academics and all intellectuals—have something to contribute, *on the condition* that they see themselves as contributing, *actively*, to the solution of real-life social problems.

The point that I have tried to make throughout this book relative to technical professionals, and that I would apply here to the community of academic philosophers, is that they should get involved in progressive, organized, public-interest efforts to solve the major social problems of our technological age. It is not, with respect to philosophers, the negative inverse of this, that (by Dewey's standards) if they do not, they are not doing "real" philosophy—and may even be contributing to the problems that block progress. It is that we have a unique opportunity today, and should do something about it.

In this conclusion I want to talk about the variety of ways in which philosophers trained in *all* traditions can make concrete contributions.

1. I do believe, though my negative remarks earlier might suggest the contrary, that analytical philosophers can make something of a contribution. I think, however, it is less likely to be to policy discussions (such as Perry's example of clarifications of the meaning of privacy) than it is to issues involving science and technology. I believe, for instance, that analytical philosophers interested in such topics as artificial intelligence may actually contribute to developments in information technology, in cognitive science, even in the understanding of brain physiology. And these developments may have social implications (not always positive).

From another perspective, applied ethicists do often seem to contribute to the solution of ethical and social problems—though under this broad heading, many things are done other than strict conceptual analysis.

2. Philosophers can also make contributions of other sorts out of their academic backgrounds. Philosophers have often contributed significantly to the systematizing and summarizing of knowledge—a task that is urgently needed to counterbalance the fragmentation of knowledge in the modern world. Indeed, philosophers may be uniquely qualified (in a sense different from Perry's) to help in encyclopedia publishing, in putting together comprehensive bibliographies and handbooks, etc.—in general, in *organizing* knowledge. Similarly, philosophers of a variety of persuasions can contribute to academic *interdisciplinary teaching* teams and programs that have the same integrative aims—here, to help students integrate their fragmented and specialized knowledge already at the undergraduate level. Even within academia, philosophers can do a public service by joining other faculty members—as Dewey did—in the fight to protect academic freedom, so often under attack today.

3. Stepping outside the ivory towers of the academy, philosophers can

make genuine and creative contributions to public commissions, to technology assessment boards, to ethics committees for various professions, and so on. In point of fact, it is often, today, religious ethicists who are invited to serve in such capacities. (In my experience, they seem to do the job quite well without having the benefit of rigorous analytical training.) Of course, applied ethicists are also invited, and some of them seem to be able to contribute significantly to public discourse, even, occasionally, by way of conceptual analysis of controversial issues.

4. The final way I .would say contemporary philosophers can contribute to the modern world is as what I would call secular preachers—advocates of *vision* in the solution of social, political, and cultural problems. Here I think of philosophers like Albert Borgmann in *Technology and the Character of Contemporary Life* and *Crossing the Postmodern Divide*.[48] Bruce Kuklick, in *The Rise of American Philosophy*,[49] maintains that this role came largely to be scorned by academic philosophers after the rise of philosophical professionalism. I believe Kuklick is, for the most part, correct. But I also believe that the small number of philosophers who still feel called upon to play this role (possibly there are larger numbers in countries other than the United States?) are not necessarily out of the philosophical mainstream, even when their preaching is covertly religious rather than completely secular. The need for vision is so great in a world of fragmented specialized knowledge that it may even be time to welcome religious thinkers back into the philosophical mainstream. Or, if that is too much to swallow, it seems correct to say that academic professionalism in philosophy would be missing an opportunity if no one today were willing to play the role of visionary secular preacher.

It is time to end my whole book with a note. I said at the outset that I think technology (to the extent one ought to speak in such generalized terms) is *not* a force that we cannot control. We must use democratic means, and the specific approach I have recommended is socially responsible activism on the part of technically-trained professionals—acting, of course, in consort with other public-spirited activists. In this chapter and in chapter two, I have even tried to draw academic philosophers into the activist circle. I also asked, early on, whether there are enough activists around to carry on the struggle to transform our technological culture in progressive ways. Possibly there are not. But possibly also there could be, and that is my hope. So in the end the book is a plea to join in the struggle—a struggle, if you will, to prove the radical critics of technology wrong. There is hope, but only if we are willing to struggle to meet the many challenges that face us in today's world.

Notes

Chapter 1. Introduction: The Crisis of Technological Culture and a Possible Solution (pages 13-27)

1. Daniel Bell, *The Cultural Contradictions of Capitalism,* 2d ed. (New York: Basic Books, 1978), 3.

2. In an early draft of this book I attempted to summarize the mountains of data supporting a claim that our age deserves a "technological" label. That now seems to me to have been a mistake. An encyclopedia would get over-filled in attempting to document such a claim. Nonetheless, one small team of scholars seems to me to have managed to do the impossible, distilling a great many discussions of modernization into a single volume in admirable fashion. I have in mind Peter Berger and his colleagues in *The Homeless Mind: Modernization and Consciousness* (New York: Random House, 1973). They call their method "phenomenological" and acknowledge Alfred Schutz as their principal mentor, with his "phenomenology of everyday life."

This same methodology had earlier served Berger (with Thomas Luckmann) well in *The Social Construction of Reality* (New York: Doubleday, 1966) —one of the most remarkable syntheses of other people's theorizing in the history of Western thought (in my opinion). Alongside Karl Marx and Sigmund Freud and Jean-Paul Sartre (as well as Schutz), a key theoretical source for Berger and Luckmann is George Herbert Mead. That is what, because of my intellectual dependence on Mead and John Dewey, drew me to Berger's books in the first place. I have, however, been less favorably impressed with some of Berger's later books, such as *Pyramids of Sacrifice* (New York: Basic Books, 1974).

3. Friedrich Nietzsche, *The Will To Power* (New York: Random House, 1967); see also Martin Heidegger's multivolume *Nietzsche* (New York: Harper, 1979-1984), especially volume 4, *Nihilism*.

4. Joseph Conrad, *The Secret Agent* (Garden City, N.Y.: Doubleday, 1953).

5. Bell, *Cultural Contradictions*, 6.

6. Ibid., 7.

7. Daniel Bell, *The Coming of Post-Industrial Society: A Venture in Social Forecasting* (New York: Basic Books, 1973).

8. Bell had earlier (1962) gained notoriety, especially among radicals, for *The End of Ideology: On the Exhaustion of Political Ideas in the Fifties: With a New Afterword* (Cambridge, Mass.: Harvard University Press, 1988).

9. Emile Durkheim, *The Rules of Sociological Method* (Glencoe, Ill.: Free Press, 1938).

10. Talcott Parsons, *Essays in Sociological Theory, Pure and Applied* (Glencoe, Ill.: Free Press, 1949).

11. Bell, *Cultural Contradictions*, 14, 18-20.

12. Ibid., 28.

13. Ibid., 30.

14. John Kenneth Galbraith, *The New Industrial State* (Boston: Houghton Mifflin, 1967), and *Economics and the Public Purpose* (Boston: Houghton Mifflin, 1973).

15. Bell, *Cultural Contradictions*, xv and elsewhere.

16. Herbert Marcuse, *One-Dimensional Man* (Boston: Beacon, 1964); *Counterrevolution and Revolt* (Boston: Beacon, 1972); and *The Aesthetic Dimension* (Boston: Beacon, 1978).

17. Morton Schoolman, *The Imaginary Witness: The Critical Theory of Herbert Marcuse* (New York: Free Press, 1980), 325.

18. Ibid.

19. Quoted in Schoolman, ibid.
20. Ibid., 346.
21. Quoted, ibid., 347.
22. Ibid.
23. Cheryl Bentsen, *Maasai Days* (New York: Summit, 1989).
24. Examples of these anecdotal reports include the Toronto *Globe and Mail* (October 26, 1989), on modern courts vs. traditional judging, and *Research/Penn State* (September 1989), on the effects of wells on settlement.
25. Jacques Ellul, *The Technological Society* (New York: Knopf, 1964), and *The Ethics of Freedom* (Grand Rapids, Mich.: Eerdmans, 1976); see Joyce M. Hanks, comp., *Jacques Ellul: A Comprehensive Bibliography* (Greenwich, Conn.: JAI Press, 1984), for a virtually complete list of works by and about Ellul.
26. D. J. Wennemann, "An Interpretation of Jacques Ellul's Dialectical Method," in *Broad and Narrow Interpretations of Philosophy of Technology*, ed. P. Durbin (Dordrecht: Kluwer, 1990), 181-192.
27. Ralph W. Sleeper, *The Necessity of Pragmatism: John Dewey's Conception of Philosophy* (New Haven, Conn.: Yale University Press, 1986), 206.
28. Ibid.
29. Larry A. Hickman, *John Dewey's Pragmatic Technology* (Bloomington and Indianapolis: Indiana University Press, 1990), 1.
30. Ibid.
31. Thomas A. Alexander, *John Dewey's Theory of Art, Experience and Nature* (Albany: State University of New York Press, 1987).
32. Ibid., xiv, citing Dewey's *Art as Experience*.
33. Ibid., xvii.
34. Ibid., 255.
35. Cornel West, *The American Evasion of Philosophy: A Genealogy of Pragmatism* (Madison: University of Wisconsin Press, 1989).
36. Richard Rorty, *Philosophy and the Mirror of Nature* (Princeton: Princeton University Press, 1979); *Consequences of Pragmatism* (Minneaplis: University of Minnesota Press, 1982); and *Contingency, Irony, and Solidarity* (New York: Cambridge University Press, 1989). Konstantin Kolenda, in *Rorty's Humanistic Pragmatism: Philosophy Democratized* (Tampa: University of South Florida Press, 1990), tries to show that Rorty is in fact interested in activism aimed at moral progress, but I do not find this explicit in Rorty's own writings. Furthermore, "moral progress" can be defined in many ways, and it is not clear that Kolenda and Rorty mean the same thing as Dewey and West.
37. See, as one advocate of such a view, Ronald Dworkin, "The Bork Nomination," *New York Review of Books*, 13 August 1987, 3-10. Dworkin there refers the reader to chapters in his three books: *Taking Rights Seriously*, chapter 5; *A Matter of Principle*, chapter 2; and *Law's Empire*, chapter 9; all three were published by Harvard University Press.
38. Deborah Johnson, in a lecture, "Do Engineers Have Social Responsibilities? Explorations in the Framework of Professional Ethics," argues that they do not. I think Johnson is wrong, and she comes much closer to the truth in "The Social/Professional Responsibility of Engineers" in *Ethical Issues Associated with Scientific and Technological Research for the Military*, ed. C. Mitcham and P. Siekevitz (New York: New York Academy of Sciences, 1989), 106-114.
39. See Bernard Bailyn, *The Ideological Origins of the American Revolution* (Cambridge, Mass.: Belknap/Harvard University Press, 1967); and Michael Kammen, *A Machine That Would Go of Itself: The Constitution in American Culture* (New York: Knopf, 1987); also, Kammen, ed., *The Origins of the American Constitution: A Documentary History* (New York: Penguin, 1987).
40. Mortimer J. Adler, *We Hold These Truths: Understanding the Ideas and Ideals of the Constitution* (New York: Macmillan, 1987); see chapters 10, 20, and 21.
41. John W. Gardner, *In Common Cause* (New York: Norton, 1972), 73.
42. Ibid.

43. Ibid., 110.
44. See Morton Grodzins and Eugene Rabinowitch, eds., *The Atomic Age: Scientists in National and World Affairs; Articles from the Bulletin of the Atomic Scientists, 1945-1962* (New York: Basic Books, 1963); Len Ackland and Steven McGuire, eds., *Assessing the Nuclear Age* (Chicago: University of Chicago Press, 1986); and Joseph Rotblat, *Scientists in Quest for Peace: A History of the Pugwash Conferences* (Cambridge, Mass.: MIT Press, 1972).
45. Rosemary Chalk, ed., *Science, Technology and Society: Emerging Relationships* (Washington, D.C.: American Association for the Advancement of Science, 1988).
46. Ibid., see 5, 6, 10, 15, 16.
47. Committee on Scientific Freedom and Responsibility, *Scientific Freedom and Responsibility* (Washington, D.C.: American Association for the Advancement of Science, 1975).
48. Rosemary Chalk, Mark Frankel, and Sallie B. Chafer, *AAAS Professional Ethics Project: Professional Ethics Activities in the Scientific and Engineering Societies* (Washington, D.C.: American Association for the Advancement of Science, 1980).
49. William Gilman, *Science: U.S.A.* (New York: Viking, 1965), 21.
50. On Ellul, see note 25, above.
51. See notes 16 and 17, above.
52. Martin Heidegger, *The Question concerning Technology and Other Essays* (New York: Harper and Row, 1977).
53. Albert Borgmann, *Technology and the Character of Contemporary Life* (Chicago: University of Chicago Press, 1984).
54. Langdon Winner, *Autonomous Technology: Technology-out-of-Control as a Theme in Political Thought* (Cambridge, Mass.: MIT Press, 1977); and *The Whale and the Reactor: A Search for Limits in an Age of High Technology* (Chicago: University of Chicago Press, 1986).
55. Samuel C. Florman, *The Existential Pleasures of Engineering* (New York: St. Martin's Press, 1976); and *Blaming Technology: The Irrational Search for Scapegoats* (New York: St. Martin's Press, 1981).
56. John Passmore, *Science and Its Critics* (New Brunswick, N.J.: Rutgers University Press, 1978).
57. Emmanuel Mesthene, *Technological Change: Its Impact on Man and Society* (Cambridge, Mass.: Harvard University Press, 1970).
58. Edward Walter, *The Immorality of Limiting Growth* (Albany: State University of New York Press, 1981).
59. See notes 1, 7, and 8, above.
60. See, for instance, Bernard Gendron's *Technology and the Human Condition* (New York: St. Martin's Press, 1977), or Josef Banka's "'Euthyphronics' and the Problem of Adapting Technical Progress to Man," in *Research in Philosophy & Technology*, vol. 2, ed. P. Durbin(Greenwich, Conn.: JAI Press, 1979); Banka has also written several books in Polish that say the same thing.
61. See the excellent discussion of STR theory in Carl Mitcham, "Philosophy of Technology," in *A Guide to the Culture of Science, Technology, and Medicine*, ed. P. Durbin (New York: Free Press, 1980, 1984), 300-303.
62. Simon Ramo, in *America's Technology Slip* (New York: Wiley, 1980), is opposed to the current regulatory system but not to all regulation.
63. See for instance, Russell Kirk, *The Conservative Mind* (Chicago: Regnery, 1953, and several later editions).

Chapter 2. Ethics as Social Problem Solving (pages 28-37)

1. George Herbert Mead, "Scientific Method and the Moral Sciences," in *Selected Writings*, ed. A. Reck (Indianapolis, Ind.: Bobbs-Merrill, 1964), 266.

2. Deborah Johnson, in a lecture, "Do Engineers Have Social Responsibilities? Explorations in the Framework of Professional Ethics," attempts to justify this (in my view) unjustifiable opinion. It is interesting that when Johnson later came to edit *Ethical Issues in Engineering* (Englewood Cliffs, N.J.: Prentice-Hall, 1991), she did not include this piece, already written at that time; instead, she included her essay, "The Social and Professional Responsibility of Engineers," in *Ethical Issues Associated with Scientific and Technological Research for the Military*, ed. C. Mitcham and P. Siekevitz (New York: New York Academy of Sciences, 1989), 106-114. That essay's conclusion—that engineers do have responsibilities that are "matters of professional judgment and professional commitment"—seems to me much more reasonable than Johnson's claim that social responsibilities are personal, not professional.

3. David L. Miller, *George Herbert Mead: Self, Language, and the World* (Austin: University of Texas Press, 1973), xxxi. See also xxxiv-xxxv.

4. Alex C. Michalos, review of Gary Bullert's *The Politics of John Dewey* in *Teaching Philosophy* 9(September 1986):282-283.

5. Gary Bullert, *The Politics of John Dewey* (Buffalo, N.Y.: Prometheus, 1983). Bullert's study has now been superseded by the much more comprehensive (and much better) study by Robert B. Westbrook, *John Dewey and American Democracy* (Ithaca, N.Y.: Cornell University Press, 1991).

6. John Dewey, *The Early Works, 1882-1889*; *The Middle Works, 1899-1924*; and *The Later Works, 1925-1952*, 35 volumes (Carbondale: Southern Illinois University Press, 1967-); there are still volumes forthcoming in the later works set. See also Jo Ann Boydston, *Guide to the Works of John Dewey* (Carbondale: Southern Illinois University Press, 1970).

7. John Herman Randall, Jr., *Aristotle* (New York: Columbia University Press, 1960), 248.

8. Willem H. Vanderburg, "Technique and Responsibility: Think Globally, Act Locally, according to Jacques Ellul," in *Technology and Responsibility*, ed. P. Durbin (Dordrecht: Reidel, 1987), 115-132.

9. Kristin S. Shrader-Frechette, *Risk Analysis and Scientific Method* (Dordrecht: Reidel, 1985); and *Science Policy, Ethics, and Economic Methodology* (Dordrecht: Reidel, 1985).

10. John Rawls, *A Theory of Justice* (Cambridge, Mass.: Harvard University Press, 1971).

11. Hans Jonas, *The Imperative of Responsibility: In Search of an Ethics for the Technological Age* (Chicago: University of Chicago Press, 1984).

12. Martin Heidegger, *Being and Time* (New York: Harper & Row, 1963); see also his *The Question Concerning Technology and Other Essays* (New York: Harper & Row, 1977).

13. Daniel Dahlstrom, "*Lebenstechnik und Essen*: Towards a Technological Ethics after Heidegger," in *Technology and Contemporary Life*, ed. P. Durbin (Dordrecht: Reidel, 1987), 145-160; see especially 157.

14. Albert Borgmann, *Technology and the Character of Contemporary Life: A Philosophical Inquiry* (Chicago: University of Chicago Press, 1984).

15. Jacques Ellul, *The Technological Society* (New York: Knopf, 1964) and *The Technological System* (New York: Continuum, 1980).

16. Gilbert Hottois, "Technoscience: Nihilistic Power versus a New Ethical Consciousness," in *Technology and Responsibility*, ed. P. Durbin (Dordrecht: Reidel, 1987), 69-84; see especially 80-81.

17. Daniel Cérézuelle, "Reflections on the Autonomy of Technology: Biotechnology, Bioethics, and Beyond," in *Technology and Contemporary Life*, ed. P. Durbin (Dordrecht: Reidel, 1987), 129-144; see especially 137 and 143; the same paper appears in French in G. Hottois, ed., *Evaluer la Technique: Aspects èthiques de la philosophie de la technique* (Paris: Vrin, 1988), 97-116.

18. Eugene Kamenka, *Marxism and Ethics* (London: Macmillan, 1969), 66-67.

19. Jozef Banka, "'Euthyphronics' and the Problem of Adapting Technical Progress to Man," in *Research in Philosophy & Technology*, vol. 2, ed. P. Durbin (Greenwich,

Conn.: JAI Press, 1979), 7. See also Bernard Gendron, *Technology and the Human Condition* (New York: St. Martin's Press, 1977), Part III.

20. See, for example, Andrew Feenberg, "Democratic Socialism and Technological Change," in *Broad and Narrow Interpretations of Philosophy of Technology*, ed. P. Durbin (Dordrecht: Kluwer, 1990), 101-123. This paper is adapted from Feenberg's *Critical Theory of Technology* (New York: Oxford University Press, 1991). Similar conclusions are defended by Carol Gould in *Rethinking Democracy* (New York: Cambridge University Press, 1988), though Gould's book was written before the recent massive changes in Communist countries.

21. David Noble, "Present Tense Technology," *Democracy: A Journal of Political Renewal and Radical Change* 3(Spring, 1983):8-24; (Summer, 1983):70-82; (Fall, 1983):71-93. The quote in the text is from 92. See also Noble's *America by Design: Science, Technology, and the Rise of Corporate Capitalism* (New York: Knopf, 1977), and *Forces of Production: A Social History of Industrial Automation* (New York: Knopf, 1984).

22. Langdon Winner, *Autonomous Technology: Technics-out-of-Control as a Theme in Political Thought* (Cambridge, Mass.: MIT Press, 1977), 107; see also 331. Winner's more mature thoughts appear in his *The Whale and the Reactor: A Search for Limits in an Age of High Technology* (Chicago: University of Chicago Press, 1986).

23. Steven L. Goldman, "Contemporary Critiques of Technology: Response and Comments," in *Contemporary Critiques of Technology*, ed. Stephen Cutcliffe (Technology Studies Working Papers; Bethlehem, Pa.: Technology Studies Resource Center, Lehigh University, 1985), 132.

24. Ralph W. Sleeper, *The Necessity of Pragmatism: John Dewey's Conception of Philosophy* (New Haven, Conn.: Yale University Press, 1986), 206-207 and 7-8.

25. Mead, "Scientific Method," 26.

26. Hans Joas, *G. H. Mead: A Contemporary Re-Examination of His Thought* (Cambridge, Mass.: MIT Press, 1985), 124.

27. Michael W. McCann, *Taking Reform Seriously: Perspectives on Public Interest Liberalism* (Ithaca, N.Y.: Cornell University Press, 1986), 262.

Chapter 3. Social Workers: A Sample of What Is to Follow (pages 38-47)

1. Pope Paul VI, *Octagesima Adveniens*, "A Call to Action," pastoral letter, 1971; reprinted in *Renewing the Earth: Catholic Documents on Peace, Justice and Liberation*, ed. D. O'Brien and T. Shannon (Garden City, N.Y.: Doubleday, 1977), 357-358.

2. See *New York Times*, 1 October 1990, A12.

3. Letty Cottin Pogrebin, *Family Politics: Love and Power on an Intimate Frontier* (New York: McGraw-Hill, 1983), 46.

4. Vance Packard, *Our Endangered Children: Growing Up in a Changing World* (Boston: Little, Brown, 1983).

5. According to the American Association for Protecting Children, as cited in Dante Cicchetti and Vicki Carlson, *Child Maltreatment: Theory and Research on the Causes and Consequences of Child Abuse and Neglect* (Cambridge, England: Cambridge University Press, 1989), xiii. Some other sources on this topic include Sander Breiner, *Slaughter of the Innocents: Child Abuse through the Ages and Today* (New York: Plenum, 1990); Robin Clark, *Encyclopedia of Child Abuse* (New York: Facts on File, 1989); David Finkelhor, *Stopping Family Violence* (Beverly Hills, Calif.: Sage, 1988); E. Clay Jorgensen, *Child Abuse: A Practical Guide for Those Who Help Others* (New York: Continuum, 1990); Oliver Tzeng and J. Jacobsen, *Sourcebook for Child Abuse and Neglect* (Springfield, Ill.: Thomas, 1988); and Michael Wald, *Protecting Abused and Neglected Children* (Stanford, Calif.: Stanford University Press, 1988).

6. See, for example, Philippe Ariès, *Centuries of Childhood: A Social History of Family Life* (New York: Knopf, 1962); and George Gerbner, Catherine J. Ross, and

Edward Zigler, eds., *Child Abuse: An Agenda for Action* (New York: Oxford University Press, 1980).

7. David Finklehor, "How Widespread Is Child Sexual Abuse?" in National Center on Child Abuse and Neglect (NCCAN), *Perspectives on Child Maltreatment in the Mid 80s* (Washington, D.C.: U.S. Department of Health and Human Services, Administration for Children, Youth and Families, Children's Bureau, 1984), 18-20.

8. Ibid., 19.

9. NCCAN, *Perspectives on Child Maltreatment*, "Overview," 28-30.

10. Peter Coolsen and Joseph Wechsler, "Community Involvement in the Prevention of Child Abuse and Neglect," in NCCAN, *Perspectives on Child Maltreatment*, 11.

11. Cheryl D. Hayes, ed., *Risking the Future: Adolescent Sexuality, Pregnancy, and Childbearing* (Washington, D.C.: National Academy of Science, 1987), vol. 1, 11-12.

12. Jeanne Warren Lindsay and Sharon Rodine, *Teen Pregnancy Challenge; Book One: Strategies for Change; Book Two: Programs for Kids* (Buena Park, Calif.: Morning Glory Press, 1989).

13. Lindsay and Rodine, *Book Two*, 197.

14. Richard B. Freeman and Harry J. Holzer, eds., *The Black Youth Employment Crisis* (Chicago: University of Chicago Press, 1986).

15. Ibid., 3.

16. Ibid.

17. Ibid., 5.

18. Ibid., 17.

19. *New York Times*, 17 July 1990, A12.

20. Brigitte Berger and Peter L. Berger, *The War over the Family: Capturing the Middle Ground* (Garden City, N.Y.: Doubleday, 1983), 202.

21. Ibid., 32-33.

22. Ibid., 33.

23. Howard Jacob Karger, "Private Practice and Social Work: A Response," *Social Work* 35(September 1990):479.

Chapter 4. Technology Educators and Reform (pages 51-59)

1. Peter L. Berger, Brigitte Berger, and Hansfield Kellner, *The Homeless Mind: Modernization and Consciousness* (New York: Random House, 1973), 40.

2. National Commission on Excellence in Education, *A Nation at Risk: The Imperative for Educational Reform* (Washington, D.C.: U.S. Department of Education, 1983). See also, among other reports, Association of American Colleges, *Integrity in the College Curriculum: A Report to the Academic Community* (Washington, D.C.: Association of American Colleges, 1985); National Governors Association, *Time for Results* (Washington, D.C.: National Governors Association, 1986); and Education Commission for the States, *Transforming the State Role in Undergraduate Education: A Time for a Different View* (Denver, Colo.: Education Commission of the States, 1986).

3. Pierre S. du Pont IV, "Real Choices: Time for Education to Be Truly Competitive," *Sunday News Journal* (Wilmington, Del.), 23 September 1990, J1 and J4.

4. Ibid.

5. E. D. Hirsch, Jr., *Cultural Literacy: What Every American Needs to Know* (Boston: Houghton Mifflin, 1987). See also Hirsch's *Dictionary of Cultural Literacy* (Boston: Houghton Mifflin, 1988), and *First Dictionary of Cultural Literacy* (Boston: Houghton Mifflin, 1989)—the last being for grades 6-8.

6. The clearest statement of Hirsch's aim appears in his response to Herbert Kohl, in "'The Primal Scene of Education': An Exchange," *The New York Review of Books* 36:6 (13 April 1989): 50-51; Kohl was replying to an article by Hirsch, "The Primal Scene of Education," *New York Review of Books* 36:3 (2 March 1989): 29-35. An account of Hirsch's Cultural Literacy Foundation at the University of Virginia can be

found in Christopher Hitchens, "Why We Don't Know What We Don't Know: Just Ask E. D. Hirsch," *New York Times Magazine*, 13 May 1990, 32, 59-60, and 62.

7. These concerns were summarized in an Education Life section in the *New York Times*, 7 January 1990, special section entitled "Science under Scrutiny," with the lead article by K. C. Cole.

8. Paul DeHart Hurd, "Historical and Philosophical Insights on Scientific Literacy," *Bulletin of Science, Technology and Society* 10:3 (1990): 133-136. That number of the *Bulletin of STS* reprints chapter 8, "The Designed World," from *Science for All Americans: A Project 2061 Report on Literacy Goals in Science, Mathematics and Technology* (Washington, D.C.: American Association for the Advancement of Science, 1989). Other materials from the Project 2061 report had been reprinted in *Bulletin STS*, 10:1 and 2.

9. *STS Reporter* 1:1(September 1988):5.

10. Ibid.

11. Hurd, "Historical and Philosophical Insights," 135-136. Steven L. Goldman had earlier argued persuasively that science courses for non-scientists (and, presumably, engineering courses for non-engineers, where such exist) take an approach exactly the opposite of the one they should take; see Goldman, "Structural Obstacles to Technology Literacy," *The Weaver*, Fall 1985, 6. Donald L. Adams, in "Science Education for Non-Majors: The Goal Is Literacy, the Method Is Separate Courses," *Bulletin of STS* 10:3(1990):125-129, bases his views on experience with one particular course taught at Babson College in Massachusetts.

12. Derek Bok, *Universities and the Future of America* (Durham, N.C.: Duke University Press, 1990).

13. Ernest L. Boyer, "Ivory Tower Has Grown Too Tall," *Sunday News Journal* (Wilmington, Del.), 9 September 1990, J1 and J4. See also Boyer, *Scholarship Reconsidered: Priorities of the Professoriate* (Princeton, N.J.: Princeton University Press, 1991). Boyer's essay is mainly a review of Bok's *Universities and the Future of America*. David Noble, in *America by Design: Science, Technology and the Rise of Corporate Capital* (New York: Knopf, 1977), demonstrates that such demands have been around for a long time—and that U.S. universities have typically responded with "reforms" along the lines demanded by the corporations.

14. *The Weaver of Information and Perspectives on Technological Literacy*, Spring 1990, 16.

15. Titles in the New Liberal Arts series include: John G. Truxal, *The Age of Electronic Messages* (New York and Cambridge, Mass.: McGraw-Hill/MIT Press, 1990); Robert Mark, *Light, Wind, and Structure: The Mystery of the Master Builders* (1990); and Joseph D. Bronzino, Vincent H. Smith, and Maurice L. Wade, *Medical Technology and Society* (1990).

16. See Truxal, *Age of Electronic Messages*, x.

17. At least this is the popular stereotype. David Noble's *America by Design* demonstrates that leading engineers and heads of technology-based corporations have long cooperated with civic leaders in an activism of a different sort.

18. James F. Rutherford, in "STS—Here Today and . . . ?," *Bulletin of Science, Technology and Society* 8:2(1988):126-127, makes some of these charges. Similar criticisms were made at a conference on technological literacy conducted in 1991 under the auspices of the Accreditation Board for Engineering and Technology and the Association of American Colleges. See Russel C. Jones, ed., *Technological Literacy Workshop: Proceedings, 6-8 May 1991* (privately printed at the University of Delaware, 1991).

19. See especially Julie Thompson Klein, *Interdisciplinarity: History, Theory, and Practice* (Detroit, Mich.: Wayne State University Press, 1990). Klein's is *the* compendium of information on the topic.

20. For an exception, see Lynn Cheney, *To Reclaim a Legacy* (Washington, D.C.: National Endowment for the Humanities, 1984). However, in Cheney's hands, this becomes a weapon in a neo-conservative attack on liberalism.

21. An outstanding example can be seen in Peter T. Marsh, ed., *Contesting the*

Boundaries of Liberal and Professional Education: The Syracuse Experiment (Syracuse, N.Y.: Syracuse University Press, 1988).

22. Julie T. Klein, "The Dialectic and Rhetoric of Disciplinarity and Interdisciplinarity," *Issues in Integrative Studies* 2(1983):35-74. This important survey is reprinted in Daryl E. Chubin et al., *Interdisciplinary Analysis and Research: Theory and Practice of Problem-Focused Research and Development* (Mt. Airy, Md.: Lomond, 1986), and the citation in the text is from p. 97 of that version. Many ideas in the balance of this chapter were first aired in a review of the Chubin book that I did for *Research in Philosophy and Technology*, vol. 9, ed. C. Mitcham (Greenwich, Conn.: JAI Press, 1989), 221-226.

23. Klein, *Interdisciplinarity*, 179-180.

24. Ibid.

25. Hans Lenk, "Toward a Pragmatic Social Philosophy of Technology and the Technological Intelligentsia," in *Research in Philosophy and Technology*, vol. 7, ed. P. Durbin (Greenwich, Conn.: JAI Press, 1984), 49.

Chapter 5. Medical Educators, Technology, and Health Reform (pages 60-70)

1. Barbara Ehrenreich and John Ehrenreich, with members of the Health-PAC, *The American Health Empire* (New York: Random House, 1970).

2. Ivan Illich, *Medical Nemesis* (New York: Pantheon, 1976).

3. See Bruce Jennings, "Grassroots Bioethics Revisited: Health Care Priorities and Community Values," *Hastings Center Report* 20(September/October, 1990):16-23. Also, Jennings, "Bioethics at the Grassroots," *Hastings Center Report* 18(June/July, 1988):special supplement.

4. "The Reform of Medical Education," *Health Affairs* 7:2 (Supplement 1988).

5. Anselm Strauss, *Social Organization of Medical Work* (Chicago: University of Chicago Press, 1985), ix. Although historian Rosemary Stevens's focus is different in *In Sickness and in Wealth* (New York: Basic Books, 1989), she too says that technology is a central and problematic feature of modern U.S. hospitals and health care; see 4ff.

6. Strauss, *Medical Work*, 3.

7. Ibid.

8. Ibid., 4.

9. Ibid., 140.

10. As suggested at a 1988 conference, "Nursing and Technology: Moving into the 21st Century" sponsored by the Food and Drug Administration, the Public Health Service, and several groups interested in health technologies. See *Health Technology* 2(September/October 1988): 212.

11. "Physicians for the Twenty-First Century," *Journal of Medical Education* 59:11, part 2, supplement (November 1984). Also available from the Association of American Medical Colleges in book form.

12. "The Reform of Medical Education," *Health Affairs* 7:2(Supplement 1988).

13. John K. Iglehart, "Medical Schools and the Public Interest: A Conversation with Robert G. Petersdorf," *Health Affairs* 7:2 (Supplement 1988): 108-120.

14. David E. Rogers, "Clinical Education and the Doctor of Tomorrow: An Agenda for Action," final chapter in *Adapting Clinical Medical Education to the Needs of Today and Tomorrow*, ed. B. Gastel and D. Rogers (New York: New York Academy of Medicine, 1988).

15. John D. Engel, Peter O. Ways, and Constance Filling, "Reflections on the General Professional Education of Physicians," *International Brain Dominance Review* 2:2 (1985): 44-51. John Engel has been very helpful to me in the writing of this chapter; he is very knowledgeable about these mainstream reform proposals and their critics—e.g., Peter Vinen-Johansen and Elianne Riska, "New Oslerians and Real Flexnerians: The Response to Threatened Professional Autonomy," *International Journal of Health Services* 21 (1991):75-108.

16. J. David Newell, "Introduction: The Nature of Medical Humanities," in *Medicine Looks at the Humanities*, ed. J. Newell and I. Gabrielson (Lanham, Md.: University Press of America, 1987), xix.

17. Eric J. Cassell, *The Place of the Humanities in Medicine* (Hastings-on-Hudson, N.Y.: Institute of Society, Ethics, and the Life Sciences, the Hastings Center, 1984), 6.

18. Joanne Trautmann, "The Wonders of Literature in Medical Education," in *The Role of the Humanities in Medical Education*, ed. D. Self (Norfolk, Va.: Bio-Medical Ethics Program, Eastern Virginia Medical School, 1978), 44.

19. Alton I. Sutnick, "Preface: Why the Humanities in Medicine?," in *Medicine Looks at the Humanities*, xiii.

20. Quoted by Patricia L. Hogg in "Medical Education Now Includes Humanities," *Pennsylvania Medicine* (February 1989), 38.

21. Something about this situation can be inferred from a special issue of *Academic Medicine* 64(December 1989):699-764. As the title, "Teaching Medical Ethics," suggests, the focus is primarily on that topic, but several broader medical humanities programs are also discussed. Another excellent source is Mary Carrington Coutts, "Teaching Ethics in the Health Care Setting, Part I: Survey of the Literature," *Kennedy Institute of Ethics Journal* (June 1991): 171-184.

22. Stephen Abrahamson, "The State of American Medical Education," *Teaching and Learning in Medicine* 2:3(1990):120-125.

23. S. W. Bloom, "The Medical School as a Social Organization: The Sources of Resistance to Change," *Medical Education* 23(1989):228.

24. Ibid. 239.

25. Kerr L. White, *The Task of Medicine: Dialogue at Wickenburg* (Menlo Park, Calif.: Henry J. Kaiser Foundation, 1988).

26. Charles E. Odegaard, *Dear Doctor: A Personal Letter to a Physician* (Menlo Park, Calif.: Kaiser Foundation, 1986).

27. White, *Task of Medicine*, 78.

28. Charles E. Odegaard, "Towards an Improved Dialogue," appendix to White, *Task of Medicine*, 109.

29. See Jacques Ellul, *The Technological Society* (New York: Knopf, 1964); Herbert Marcuse, *One-Dimensional Man* (Boston: Beacon, 1964) and David F. Noble, *America by Design: Science, Technology, and the Rise of Corporate Capitalism* (New York: Knopf, 1977). None of these so-called antitechnologists focuses specifically on technology in medicine, but there is every reason to believe that they would endorse at least the major theses of the radical health critics referenced in the next note.

30. See Ehrenreich and Erhenreich, *American Health Empire*, and Illich, *Medical Nemesis*, as well as Vicente Navarro, *Medicine under Capitalism* (New York: Prodist, 1976), and *Crisis, Health, and Medicine: A Social Critique* (New York: Tavistock, 1986).

31. Bruce Jennings, "Democracy and Justice in Health Policy," *Hastings Center Report* 20 (September/October 1990): 23.

Chapter 6. Media Professionals and Politics (pages 71-84)

1. Peter L. Berger, Brigitte Berger, and Hansfried Kellner, *The Homeless Mind: Modernization and Consciousness* (New York: Random House, 1973), 40.

2. David Noble, *America by Design: Science, Technology, and the Rise of Corporate Capitalism* (New York: Knopf, 1977).

3. *New York Times*, 28 October 1990, 1.

4. Joel L. Swerdlow, ed., *Media Technology and the Vote: A Sourcebook* (Washington, D.C.: Annenberg; and Boulder, Colo.: Westview, 1988). See also Douglas Kellner, *Television and the Crisis of Democracy* (Boulder, Colo.: Westview, 1990).

5. Paul Taylor, *See How They Run: Electing the President in an Age of Mediaocracy* (New York: Knopf, 1990).

6. Roger Simon, *Road Show: In America, Anyone Can Become President. It's One of the Risks We Take* (New York: Farrar, Straus, & Giroux, 1990).

7. Nicholas Lemann, "How the Spin Doctors Operate," review of Taylor, *See How They Run*, and Simon, *Road Show*; *New York Times Book Review*, 30 September 1990, 13.

8. Lemann, "Spin Doctors," 13.

9. David Halberstam, *The Powers That Be* (New York: Knopf, 1979).

10. Ibid., 728.

11. Herbert J. Gans, *Deciding What's News: A Study of CBS Evening News, NBC Nightly News, Newsweek, & Time* (New York: Random House, 1979).

12. Ibid., xiv.

13. Ibid.

14. Ibid., 212.

15. Ibid.

16. Roy Reed, "From the Campuses: Adventures in Publishing," *New York Times Book Review*, 24 September 1989, 1 and 61-62.

17. Ibid., 61.

18. Herbert I. Schiller, *Mass Communications and American Empire* (New York: Kelley, 1969); *Who Knows: Information in the Age of the Fortune 500* (Norwood, N.J.: Ablex, 1981); *Information and the Crisis Economy* (Norwood, N.J.: Ablex, 1984); and *Culture Inc.: The Corporate Takeover of Public Expression* (New York: Oxford University Press, 1989).

19. Schiller, *Culture, Inc.*, 3-4.

20. Ibid., 29.

21. Ibid., 43-44.

22. Ibid., 156.

23. Ibid., 174.

24. John W. Gardner, *In Common Cause* (New York: Norton, 1972).

25. Ibid., 65.

26. Ibid., 89.

27. Jeffrey M. Berry, *Lobbying for the People: The Political Behavior of Public Interest Groups* (Princeton, N.J.: Princeton University Press, 1977), 245.

28. Ibid., 247. Berry mentions 25 studies in all by Nader's groups before 1972.

29. Ibid., 248-249.

30. Michael W. McCann, *Taking Reform Seriously: Perspectives on Public Interest Liberalism* (Ithaca, N.Y.: Cornell University Press, 1986), 52.

31. Ibid., 52-53.

32. Donald Ross, *A Public Citizen's Action Manual* (New York: Grossman, 1973). In a bibliography that includes a section on primary sources, McCann, *Taking Reform Seriously*, lists this along with Ralph Nader and Donald Ross, *Action for Change: A Student Manual for Public Interest Organizing* (New York: Grossman, 1971), and half a dozen similar action manuals.

33. *Media Access Project 1979 Annual Report*.

34. *Media Access Project 1980 Annual Report*, 34-36.

35. Gans, *Deciding What's News*, 312.

36. Jacques Ellul, "The Latest Developments in Technology and the Philosophy of the Absurd," in *Research in Philosophy and Technology*, vol. 7, ed. P. Durbin (Greenwich, Conn.: JAI Press, 1984), 89.

37. Kathryn C. Montgomery, in *Target, Prime Time: Advocacy Groups and the Struggle over Entertainment Television* (New York: Oxford University Press, 1989), discusses a broad range of groups protesting the contents of TV programs—women, blacks, Hispanics, gays, pro- and anti-abortion groups, parents, religious groups, etc.—but focuses only at the end on the dangers of a market-driven TV industry. Obviously, her book, with its special focus, would say nothing about corporate dominance of TV and the rest of the media, let alone anything about the influence of

TV on politics.

Chapter 7. Biotechnologists, Scientific Activism, and Philosophical Bridges (pages 87-96)

1. Jeremy Rifkin, *Algeny* (New York: Viking, 1983), 238.
2. Neil A Holtzman, *Proceed with Caution: Predicting Genetic Risks in the Recombinant DNA Era* (Baltimore, Md.: Johns Hopkins University Press, 1989), 247.
3. Paul DeForest, et al., eds., *Biotechnology: Professional Issues and Social Concerns* (Washington, D.C.: Committee on Scientific Freedom and Responsibility, American Association for the Advancement of Science, 1988), ii-iii.
4. Ibid., iv.
5. "Agricultural Biotechnology Issues," *Agriculture and Human Values* 5:3 (Summer 1988): entire issue.
6. Gary Comstock, "The Case against bGH," *Agriculture and Human Values* 5:3(Summer 1988):49.
7. Mark Sagoff, "Biotechnology and the Environment: What Is at Risk?" *Agriculture and Human Values* 5:3(Summer 1988):31-32.
8. J. Sousa Silva, "The Contradictions of the Biorevolution for the Development of Agriculture in the Third World: Biotechnology and Capitalist Interest," *Agriculture and Human Values* 5:3(Summer 1988):68.
9. Rifkin, *Algeny*, 231-233, 236, and 237-238.
10. Hans Jonas, "Toward a Philosophy of Technology," *The Hastings Center Report* 9 (February 1979):34-43. See also Jonas's earlier essay, "Biological Engineering: A Preview," in his *Philosophical Essays* (Englewood Cliffs, N.J.: Prentice-Hall, 1974), 141-167; a later essay, "Ethics and Biogenic Art," *Social Research* 52:3 (1985): 491-504; and *The Imperative of Responsibility: In Search of an Ethics for the Technological Age* (Chicago: University of Chicago Press, 1984).
11. President's Commission for the Study of Ethical Problems in Medicine and Biomedical and Behavioral Research, *Splicing Life* (Washington, D.C.: U.S. Government Printing Office, 1983).
12. Sheldon Krimsky, *Genetic Alchemy: The Social History of the Recombinant DNA Controversy* (Cambridge, Mass.: MIT Press, 1982), 17. More recently Krimsky has continued his philosophical/historical account of biotechnology with *Biotechnics and Society: The Rise of Industrial Genetics* (New York: Preaeger, 1991).
13. Krimsky, *Genetic Alchemy*, 21-22.
14. Ibid., 23.
15. Avner Cohen and Marcelo Dascal, eds., *The Institution of Philosophy: A Discipline in Crisis?* (LaSalle, Ill.: Open Court, 1989).
16. Bruce Kuklick, *The Rise of American Philosophy: Cambridge, Massachusetts, 1860-1930* (New Haven, Conn.: Yale University Press, 1977), 291 and 452. Kuklick's view is supported by Daniel Wilson's *Science, Community, and the Transformation of American Philosophy, 1860-1930* (Chicago: University of Chicago Press, 1990).
17. George Herbert Mead, *Selected Writings*, ed. A. Reck (Indianapolis, Ind.: Bobbs-Merrill, 1964), xxv.
18. Gary Bullert, *The Politics of John Dewey* (Buffalo, N.Y.: Prometheus, 1983), 9. Bullert's interpretation of Dewey as a public philosopher has been deepened and corrected by Robert B. Westbrook in *John Dewey and American Democracy* (Ithaca, N.Y.: Cornell University Press, 1991).
19. Bullert, *Politics of Dewey*, 11-12.
20. For examples, see Paul Ramsey, *Fabricated Man: The Ethics of Genetic Control* (New Haven, Conn.: Yale University Press, 1970); Joseph Fletcher, *The Ethics of Genetic Control: Ending Reproductive Roulette* (Garden City, N.Y.: Doubleday, 1974); Jonathan D. Moreno, "Private Genes and Public Ethics," *Hastings Center Report* 13 (October 1983); Jonathan Glover, *What Sort of People Should There Be?* (New York:

Penguin, 1984); John C. Fletcher, "Ethical Issues in and beyond Prospective Clinical Trials of Human Gene Therapy," *Journal of Medicine and Philosophy* 10 (August 1985): 293-309; also, David A. Jackson and Stephen C. Stich, eds., *The Recombinant DNA Debate* (Englewood Cliffs, N.J.: Prentice-Hall, 1979), and William Walters and Peter Singer, eds., *Test-Tube Babies: A Guide to Moral Questions, Present Techniques, and Future Possibilities* (Melbourne: Oxford University Press, 1982).

21. Michael D. Bayles, *Professional Ethics*, 2d ed. (Belmont, Calif.: Wadsworth, 1989).

22. Joan C. Callahan ed., *Ethical Issues in Professional Life* (New York: Oxford University Press, 1988).

Chapter 8. Computer Professionals and Civil Liberties (pages 97-104)

1. Warren Freedman, in *The Right of Privacy in the Computer Age* (Westport, Conn.: Greenwood Press/Quorum Books, 1987), spells out the details of privacy law and what individuals can do to protect themselves. A very interesting summary of how electronic and computerized data collecting, quite innocent at the beginning, can snowball toward a "dossier society" is provided by Peter H. Lewis in a column on personal computers in the *New York Times*, 2 October 1990, C8. Readers might also enjoy *Privacy Journal: An Independent Monthly on Privacy in a Computer Age*, a newsletter published by Robert Ellis Smith in Washington, D.C. My focus here is not on individuals but on collective action.

2. Deborah G. Johnson, *Computer Ethics* (Englewood Cliffs, N.J.: Prentice-Hall, 1985), 3.

3. Michael Bayles, *Professional Ethics*, 2d ed. (Belmont, Calif.: Wadsworth, 1989), x.

4. Office of Technology Assessment, *Electronic Surveillance and Civil Liberties* (Washington, D.C.: U.S. Congress Office of Technology Assessment, October 1985), iii.

5. Ibid., 4.

6. OTA, *Management, Security, and Congressional Oversight* (Washington, D.C.: U.S. Congress Office of Technology Assessment, February 1986), iii.

7. OTA, *Electronic Record Systems and Individual Privacy* (Washington, D.C.: U.S. Congress Office of Technology Assessment, June 1986), 7. Later studies were published in this series.

8. OTA, *Electronic Record Systems*, 7.

9. Ibid.

10. Ibid.

11. OTA, *Electronic Surveillance and Civil Liberties*, 5

12. See David F. Linowes, *Privacy in America: Is Your Private Life in the Public Eye?* (Urbana: University of Illinois Press, 1989), chapter 5, especially 75-80. Linowes was chairperson of the U.S. Privacy Protection Commission from 1975 to 1977 and continued to serve Congress, the President, and many federal agencies as an adviser on privacy issues for over a decade afterward.

13. Johnson, *Computer Ethics*, 4.

14. Ibid., 5.

15. Carl Mitcham, in "Computer Ethos, Computer Ethics," in *Research in Philosophy and Technology*, vol. 8, ed. P. Durbin (Greenwich, Conn.: JAI Press, 1985), 267-280, reviews about a dozen recent books on or related to computer ethics, including two by Johnson.

16. Sherry Turkle, *The Second Self: Computers and the Human Spirit* (New York: Simon & Schuster, 1984).

17. See Bayles, *Professional Ethics*, for the best introduction to what these responsibilities are. Unfortunately, Bayles is little more outspoken than Johnson on the issue of getting professionals to do something about shouldering their responsibilities.

18. Marc Rotenberg, Mary J. Culnan, and Ronni Rosenberg, "Summary Statement on Computer Privacy and H.R. 3669, the Data Protection Act of 1990," before the U.S. House of Representatives Subcommittee on Government Information, Justice, and Agriculture (Committee on Government Operations), 16 May 1990, 7. Rotenberg was speaking for CPSR.

19. Joseph Weizenbaum, *Computer Power and Human Reason: From Calculation to Judgment* (San Francisco, Calif.: Freeman, 1976).

20. See *Physics Today*, 38(November 1985):82.

21. See *Bulletin of the Atomic Scientists*, 42(January 1986):28.

22. See Philip M. Boffey, *Claiming the Heavens: The New York Times Complete Guide to the Star Wars Debate* (New York: Times Books, 1988), 162.

23. See Herbert I. Schiller, *Information and the Crisis Economy* (Norwood, N.J.: Ablex, 1984); *Who Knows: Information in the Age of the Fortune 500* (Norwood, N.J.: Ablex, 1981); and the much earlier *Mass Communications and American Empire* (New York: Kelley, 1969).

Chapter 9. Nuclear Experts and Activism (pages 105-117)

1. Samuel McCracken, *The War against the Atom* (New York: Basic Books, 1982), chapter 1; and Bernard L. Cohen, *Before It's Too Late: A Scientist's Case for Nuclear Energy* (New York: Plenum, 1983), chapters 1 and 3.

2. Richard G. Hewlett and Oscar E. Anderson, Jr., *A History of the United States Atomic Energy Commission*, vol. 1: *The New World, 1939-1946* (University Park: Pennsylvania State University, 1969).

3. John M. Byrne, Steven M. Hoffman, and Cecilia Martinez, "Technological Politics in the Nuclear Age," *Bulletin of Science, Technology, and Society* 8:6(1988):580.

4. Kristin S. Shrader-Frechette, "The Plutonium Economy: Technological Links and Epistemological Problems," in *Research in Philosophy and Technology*, vol. 8, ed. P. Durbin (Greenwich, Conn.: JAI Press, 1985), 191.

5. Ibid., 141.

6. Ibid.

7. Joseph G. Morone and Edward J. Woodhouse, *The Demise of Nuclear Energy?: Lessons for Democratic Control of Technology* (New Haven, Conn.: Yale University Press, 1989), 33.

8. This is mostly based on her earlier book, *Nuclear Power and Public Policy: The Social and Ethical Problems of Fission Technology* (Dordrecht: Reidel, 1980).

9. See Dorothy Nelkin, *Technological Decisions and Democracy: European Experiments in Public Participation* (Beverly Hills, Calif.: Sage, 1977); also Nelkin and Michael Pollak, *The Atom Besieged: Extraparliamentary Dissent in France and Germany* (Cambridge, Mass.: MIT Press, 1981).

10. Dorothy Nelkin, *Nuclear Power and Its Critics* (Ithaca, N.Y.: Cornell University Press, 1971).

11. Nelkin, "Nuclear Power and Its Critics: A Siting Dispute," in Nelkin, ed., *Controversy: Politics of Technical Decisions*, 2d ed. (Beverly Hills, Calif.: Sage, 1984), 52.

12. Ibid., 54.

13. Byrne, Hoffman, and Martinez, "Technological Politics," 590.

14. Ibid.

15. Ibid., 589.

16. Ibid., 590.

17. Morone and Woodhouse, *Demise of Nuclear Energy*, 129.

18. Ibid.

19. Ibid., 155.

20. Ibid.

21. Carl Mitcham, "The Spectrum of Ethical Issues Associated with the Military Support of Science and Technology," in *Ethical Issues Associated with Scientific and Technological Research for the Military*, ed. Mitcham and P. Siekevitz (Annals of the New York Academy of Sciences, vol. 577; New York: New York Academy of Sciences, 1989), 1-9.
22. Daniel J. Kevles, *The Physicists: The History of a Scientific Community in Modern America* (New York: Knopf, 1978), 112.
23. Ibid., 287.
24. Ibid.
25. Ibid., 320.
26. Douglas MacLean, "Masters of War: The Moral Arguments and the Traditions of Ethics," in Mitcham and Siekevitz, *Research for the Military*, 38.
27. Ibid.
28. Rosemary Chalk, "Drawing the Line: An Examination of Conscientious Objection in Science," in Mitcham and Siekevitz, *Research for the Military*, 61.
29. Ibid., 63.
30. Ibid.
31. Ibid., 67.
32. Stephen H. Unger, "Engineering Ethics and the Question of Whether to Work on Military Projects," in Mitcham and Siekevitz, *Research for the Military*, 213.
33. Ibid., 215.
34. Frances B. McCrea and Gerald E. Markle, *Minutes to Midnight: Nuclear Weapons Protest in America* (Newbury Park, Calif.: Sage, 1989), 161-162.
35. Ibid.
36. Ibid., 144.
37. Ibid., 156.
38. Ibid., 158.
39. Ibid., 157.
40. Paul T. Durbin, "The Moral Arguments and the Philosophy of Science and Technology," in Mitcham and Siekevitz, *Research for the Military*, 47-50; the remainder of this chapter borrows heavily from that essay.

Chapter 10. Environmentalists and Environmental Ethics (pages 118-126)

1. Eugene Hargrove, "On Reading Environmental Ethics," *Environmental Ethics* 6 (1984): 291.
2. John Lemons, "Comment: A Reply to 'On Reading Environmental Ethics,' *Environmental Ethics* 7 (1985):185-188.
3. Some examples: Robin Attfield, *The Ethics of Environmental Concern* (New York: Columbia University Press, 1983); H. J. McCloskey, *Ecological Ethics and Politics* (Totowa, N.J.: Rowman & Littlefield, 1983); John Passmore, *Man's Responsibility for Nature* (London: Duckworth, 1980); Holmes Rolston, III, *Environmental Ethics* (Philadelphia, Pa.: Temple University Press, 1988); Mark Sagoff, *The Economy of the Earth* (New York: Cambridge University Press, 1988); Christopher D. Stone, *Earth and Other Ethics: The Case for Moral Pluralism* (New York: Harper & Row, 1987); Paul W. Taylor, *Respect for Nature: A Theory of Environmental Ethics* (Princeton, N.J.: Princeton University Press, 1986). Anthologies include: Ian Barbour, ed., *Western Man and Environmental Ethics* (Reading, Mass.: Addison-Wesley, 1973); William T. Blackstone, ed., *Philosophy and Environmental Crisis* (Athens: University of Georgia Press, 1974); Ernest Partridge, ed., *Responsibilities to Future Generations: Environmental Ethics* (Buffalo, N.Y.: Prometheus, 1981); Tom Regan, ed., Earthbound: *New Introductory Essays in Environmental Ethics* (Philadelphia, Pa.: Temple University Press, 1984); Donald Scherer and Thomas Attig, eds., *Ethics and the Environment* (Englewood Cliffs, N.J.: Prentice-Hall, 1983); Kristin Shrader-Frechette, *Environmental Ethics* (Pacific Grove, Calif.: Boxwood Press, 1981); and Donald van de Veer and

Christine Pierce, eds., *People, Penguins, and Plastic Trees: Basic Issues in Environmental Ethics* (Belmont, Calif.: Wadsworth, 1986).

4. John Dewey, *Reconstruction in Philosophy* (Boston: Beacon, 1948), and *Liberalism and Social Action* (New York: Putnam, 1935).

5. Ralph W. Sleeper, *The Necessity of Pragmatism: John Dewey's Conception of Philosophy* (New Haven, Conn.: Yale University Press, 1986).

6. Anna Bramwell, *Ecology in the Twentieth Century: A History* (New Haven, Conn.: Yale University Press, 1989).

7. Robert C. Paehlke, *Environmentalism and the Future of Progressive Politics* (New Haven, Conn.: Yale University Press, 1989), 14-22. Paehlke cites, as the best histories, Samuel P. Hays, *Conservation and the Gospel of Efficiency: The Progressive Conservation Movement, 1890-1920* (Cambridge, Mass.: Harvard University Press, 1959); Roderick Nash, *Wilderness and the American Mind* (New Haven, Conn.: Yale University Press, 1967); and Donald Worster, *Nature's Economy: A History of Ecological Ideas* (Cambridge: Cambridge University Press, 1985). See also Donald Worster, ed., *American Environmentalism: The Formative Period, 1860-1915* (New York: Wiley, 1973).

8. George Perkins Marsh, *Man and Nature: Physical Geography as Modified by Human Action* (Cambridge, Mass.: Harvard University Press, 1965).

9. See Franklin Russell, "The Vermont Prophet: George Perkins Marsh," *Horizon* 10 (1968):17.

10. See John Muir, *The Wilderness World of John Muir*, ed. E. Teale (Boston: Houghton Mifflin, 1954), and Aldo Leopold, *A Sand County Almanac* (New York: Ballantine, 1970).

11. Rachel Carson, *Silent Spring* (Boston: Houghton Mifflin, 1962). First appeared in *The New Yorker* in 1960.

12. Paehlke, *Environmentalism*, 21.

13. Frank Graham, Jr., *Since Silent Spring* (Boston: Houghton Mifflin, 1970).

14. Ibid., 22.

15. Barry Commoner, *Science and Survival* (New York: Viking, 1963), and *The Closing Circle: Nature, Man and Technology* (New York: Knopf, 1971).

16. Paul R. Ehrlich, *The Population Bomb* (New York: Ballantine, 1968).

17. See Paul R. Ehrlich and Anne H. Ehrlich, *Extinction: The Causes and Consequences of the Disappearance of Species* (New York: Random House, 1981); Paul R. Ehrlich, Anne H. Ehrlich, and John P. Holdren, *Ecoscience: Population, Resources, Environment* (San Francisco: Freeman, 1977); also, Paul R. Ehrlich, *The Cold and the Dark: The World After Nuclear War* (New York: Norton, 1984).

18. Dennis Meadows, Donella Meadows, et al., *The Limits to Growth* (New York: Universe, 1972).

19. See, for instance, H. S. D. Cole, ed., *Thinking about the Future: A Critique of the Limits to Growth* (London: Chatto & Windus, 1973), and Robert McCutcheon, *Limits of a Modern World: A Study of the Limits to Growth Debate* (London: Butterworths, 1979).

20. World Commission on Environment and Development, *Our Common Future* (New York: Oxford University Press, 1987).

21. Rae Goodell, *The Visible Scientists* (Boston: Little, Brown, 1977).

22. Jacqueline Cramer, "Societal Role of Dutch Fresh-Water Ecologists in Environmental Policies," in *Technology and Responsibility*, ed. P. Durbin (Dordrecht, The Netherlands: Reidel, 1987).

23. Cramer, "Societal Role," 271 and 272.

24. See Carol E. Hoffecker, *Delaware: A Bicentennial History* (New York: Norton, 1977), 63-64 and 206-207.

25. An example: Council on Environmental Quality, *The Delaware River Basin: An Environmental Assessment of Three Centuries of Change* (Washington, D.C.: U.S. Government Printing Office, 1975).

26. Peterson's early citizen activism was associated with the First Unitarian Church in Wilmington, Delaware. An appeal to his old friends there to continue their support

is described in *The World: Journal of the Unitarian Universalist Association* 4 (March/April 1990): 17-18. Other appeals have appeared regularly, each time a new threat to the Coastal Zone Act appeared on the horizon, in the Wilmington *News-Journal.*

27. Paehlke, *Environmentalism and the Future of Progressive Politics* (see note 7, above).

28. Ibid., 3.

29. Ibid.

30. Ibid., 276.

31. Ibid.

32. Ibid., 277.

33. Lester R. Brown, Editor's Page, *World Watch* 3, no. 1 (January-February 1990):2. See also *State of the World 1992* (New York: Norton, 1992).

34. See note 20, above.

35. Final Report of the Seventy-Seventh American Assembly (Columbia University): *Preserving the Global Environment: The Challenge of Shared Leadership.* See the set of papers under this title edited by Jessica Tuchman Mathews (New York: Norton, 1991).

36. Lemons, "Comment," 188.

37. This chapter appears as an article in *Research in Philosophy and Technology*, vol. 12, ed. Frederick Ferré (Greenwich, Conn.: JAI Press, 1992), 107-117. It is countered there by George Allan's "Environmental Philosophizing and Environmental Activism." In a nutshell, Allan says I distort Dewey's call for activism and ignore the need for philosophers to provide a speculative vision that will allow us to distinguish between good and bad activism. I think Allan is right that there is a need for clear-thinking and imaginative environmental philosophers, but I refuse to back down from my claim that they will only be effective if they join forces with progressive social activists so that, together, we can mount an effective attack on specific environmental (and related) problems.

Chapter 11. A Neo-Marxist Challenge to Progressive Activism (pages 129-139)

1. Peter L. Berger and Thomas Luckmann, *The Social Construction of Reality* (Garden City, N.Y.: Doubleday, 1966), 60.

2. Bernard Gendron, "Reply: Growth, Power, and the Imperatives of Technology," in *Research in Philosophy & Technology*, vol. 3, ed. P. Durbin (Greenwich, Conn.: JAI Press, 1980), 102-116, esp. 111-115.

3. Jacques Ellul, "The Techological Order," in *The Technological Order*, ed. C. Stover (Detroit: Wayne State University Press, 1963), 10.

4. See Paul T. Durbin, "Toward a Social Philosophy of Technology," in *Research in Philosophy & Technology*, vol. 1, ed. P. Durbin, (Greenwich, Conn.: JAI Press, 1978), 67-97, esp. 87 and 97; and "Reviews of Bernard Gendron, *Technology and the Human Condition*, I" in *Research in Philosophy & Technology*, vol. 3, ed. P. Durbin (Greenwich, Conn.: JAI Press, 1980), 77-87.

5. Willis H. Truitt, "Values in Science," in *Research in Philosophy & Technology*, vol. 1, 119-130, esp. 127-130.

6. Gendron, "Reply," 111-115.

7. Albert Borgmann, *Technology and the Character of Contemporary Life* (Chicago: University of Chicago Press, 1984), 82-85.

8. Manfred Stanley, "Symposium on Albert Borgmann, II. A Critical Appreciation," in *Technology and Contemporary Life*, ed. P. Durbin (Dordrecht: Reidel, 1988), 17-21.

9. Stanley lists, as the best example, David Harvey, but also (among others) Louis Althusser, William Connolly, Maurice Godelier, Herbert Guttman, Jurgen Habermas, James O'Connor, E. P. Thompson, and Robert Paul Wolff. See also David McLellan,

Marxism after Marx: An Introduction (New York: Harper & Row, 1979).

10. See Karl Marx, *Selected Writings*, ed. D. McLellan (Oxford: Oxford University Press, 1977), 93.

11. As I have said several times, the best recent interpretations of Dewey—R. W. Sleeper, *The Necessity of Pragmatism: John Dewey's Conception of Philosophy* (New Haven, Conn.: Yale University Press, 1986); Larry Hickman, *Dewey's Pragmatic Technology* (Bloomington: Indiana University Press, 1990); and Cornel West, *The American Evasion of Philosophy: A Genealogy of Pragmatism* (Madison: University of Wisconsin Press, 1989)—all present him as fundamentally meliorist. Sleeper documents Dewey's meliorism; Hickman interprets it as a critique of technological culture; and West relates pragmatism to new movements for social justice in a multicultural historical context. In my view, the best recent treatment of Mead is Hans Joas, *G. H. Mead: A Contemporary Re-Examination of His Thought* (Cambridge, Mass.: MIT Press, 1985). Joas draws out the (non-Communist) socialist implications of Mead's social psychology in an admirable fashion.

12. See, for example, Andrew Feenberg, "Democratic Socialism and Technological Change," in *Broad and Narrow Interpretations of Philosophy of Technology*, ed. P. Durbin (Dordrecht: Kluwer, 1990), 101-123. Similar conclusions are defended by Carol Gould in *Rethinking Democracy* (New York: Cambridge University Press, 1988), though her book was written before recent changes in Eastern Europe.

13. See chapter 2, above.

14. Eugene D. Genovese, *Roll, Jordan, Roll: The World That the Slaves Made* (New York: Pantheon, 1974), 86.

15. Berger and Luckmann, *Social Construction*, 59; see also 60.

16. Ibid., 125-127.

17. Herbert Marcuse, *One-Dimensional Man* (Boston: Beacon, 1964).

18. David Noble, *America by Design: Science, Technology and the Rise of Corporate Capitalism* (New York: Knopf, 1977).

19. Peter Berger, Brigitte Berger, and Hansfield Kellner, *The Homeless Mind: Modernization and Consciousness* (New York: Random House, 1973).

20. Allen E. Buchanan, *Marx and Justice: The Radical Critique of Liberalism* (Totowa, N.J.: Rowman and Allanheld, 1981).

21. Gary Bullert, *The Politics of John Dewey* (Buffalo, N.Y.: Prometheus, 1983).

22. Robert B. Westbrook, *John Dewey and American Democracy* (Ithaca, N.Y.: Cornell University Press, 1991).

23. See Larry Hickman, "Doing and Making in a Democracy: Dewey's Experience of Technology," and Edmund Byrne, "Workplace Democracy for Teachers: John Dewey's Contribution," both in *Philosophy of Technology: Practical, Historical, and Other Perspectives*, ed. P. Durbin (Dordrecht: Kluwer, 1989). Also, Hickman's *Dewey's Pragmatic Technology*, and Byrne's *Work, Inc.* (Philadelphia, Pa.: Temple University Press, 1990). Byrne's book is discussed at length in chapter 13, below.

24. See George Herbert Mead, *Selected Writings*, ed. A. Reck (Indianapolis: Bobbs-Merrill, 1964), xxxiii-xxxvi. Some aspects of this part of Mead's life are available also in Joas's intellectual biography (see note 11, above), but we badly need an adequate biography of Mead.

25. James Campbell, "George Herbert Mead on Social Fusion and the Social Critic," and William M. O'Meara, "Marx and Mead on the Social Nature of Rationality and Freedom." The papers from the Frontiers in American Philosophy Conference, held at Texas A&M University in June 1988, are supposed to be published in a proceedings volume, but details are not yet clear.

26. Michael W. McCann, *Taking Reform Seriously: Perspectives on Public Interest Liberalism* (Ithaca, N.Y.: Cornell University Press, 1986). Morton Schoolman, in "Liberalism's Ambiguous Legacy: Individuality and Technological Constraints," in *Research in Philosophy & Technology*, vol. 7, ed. P. Durbin (Greenwich, Conn.: JAI Press, 1984), 229-252, bases his optimism about "radicalized liberal politics" on more speculative—but still interesting—foundations.

27. See Bullert's *Politics of John Dewey*, 130ff., as well as Dewey's works cited

there.

Chapter 12. Autonomous Technology Theorists: A Second Challenge (pages 140-149)

1. Martin Heidegger, *Nietzsche, Volume IV: Nihilism* (New York: Harper & Row, 1982), 196. See also Heidegger's *The Question Concerning Technology and Other Essays* (New York: Harper & Row, 1977).
2. Jacques Ellul, "The Technological Order," in *The Technological Order*, ed. C. Stover (Detroit, Mich.: Wayne State University Press, 1963), 10-11. See also Ellul's *The Technological Society* (New York: Knopf, 1964) and *The Technological System* (New York: Continuum, 1980).
3. John Dewey, *Reconstruction in Philosophy* (Boston: Beacon Press, 1948).
4. Langdon Winner, *Autonomous Technology: Technics-out-of-Control as a Theme in Political Thought* (Cambridge, Mass.: MIT Press, 1977).
5. Ibid., 8 and 11.
6. Ibid., x and 2.
7. Ibid., 42-43.
8. Ibid., 226.
9. Ibid., 107, 171, and 276-277.
10. Ibid., 277.
11. Ibid., 331.
12. Paul T. Durbin, "Some Contributions toward an Assessment of the Place of Science and Technology in Contemporary Culture," reviews of Winner's *Autonomous Technology*, David Noble's *America by Design*, and Daniel Kevles's *The Physicists*; *Social Indicators Research* 7(1980):495-504.
13. Langdon Winner, *The Whale and the Reactor: A Search for Limits in an Age of High Technology* (Chicago: University of Chicago Press, 1986).
14. This paper of Winner's had previously appeared in several forms in several places. For one earlier version, see Winner, "*Techne* and *Politeia*: The Technical Constitution of Society," in *Philosophy and Technology*, ed. P. Durbin and F. Rapp (Dordrecht: Reidel, 1983), 97-111.
15. Winner, *Whale and Reactor*, 44-46.
16. Ibid., 175.
17. Ibid.
18. Ibid., 177.
19. Albert Borgmann, *Technology and the Character of Contemporary Life: A Philosophical Inquiry* (Chicago: University of Chicago Press, 1984).
20. Paul T. Durbin, review of Albert Borgmann, *Technology and the Character of Contemporary Life; Man and World* 21(1988):231-235.
21. Ibid., 201-202 and 191.
22. Ibid., 201.
23. Ibid., 9 and 11-12.
24. Ibid., 6-7.
25. Ibid., 246.
26. Ibid., 247.
27. Dewey, *Reconstruction in Philosophy*, xi-xii.

Chapter 13. Workers: A Response to Radical Critics (pages 150-157)

1. I focus here on these issues as presented in Edmund Byrne, *Work, Inc.: A Philosophical Inquiry* (Philadelphia, Pa.: Temple University Press, 1990). Much of the material in this chapter appears as a review of Byrne's book in *Research in Philosophy*

and Technology, vol. 12, ed. F. Ferré (Greenwich, Conn.: JAI Press, 1992), 356-360.

2. Andrew Feenberg, "Democratic Socialism and Technological Change," in *Broad and Narrow Interpretations of Philosophy of Technology*, ed. P. Durbin (Dordrecht: Kluwer, 1990), 101-123.

3. Richard E. Sclove, "The Nuts and Bolts of Democracy: Toward a Democratic Politics of Technological Design," in *Critical Perspectives on Nonacademic Science and Engineering*, ed. P. Durbin (Bethlehem, Pa.: Lehigh University Press, 1991), 239-262.

4. *The* basic reference these days (and the one Byrne keys his work to) is John Rawls, *A Theory of Justice* (Cambridge, Mass.: Belknap/Harvard University Press, 1971). For reactions to Rawls, see Norman Daniels, ed., *Reading Rawls* (New York: Basic Books, 1974), Brian Barry, *The Liberal Theory of Justice* (Oxford: Clarendon/Oxford University Press, 1973), and Robert Paul Wolff, *Understanding Rawls* (Princeton, N.J.: Princeton University Press, 1977). For other views, see Robert Nozick, *Anarchy, State, and Utopia* (New York: Basic Books, 1974); Jan Narveson, *The Libertarian Idea* (Philadelphia, Pa.: Temple University Press, 1989); Michael J. Sandel, *Liberalism and the Limits of Justice* (Cambridge: Cambridge University Press, 1982); and especially Alan Brown, *Modern Political Philosophy: Theories of the Just Society* (New York: Viking Penguin, 1986). For more traditional (but still relevant) views, one should consult *The Great Ideas: A Syntopicon of Great Books of the Western World*, ed. M. Adler (Chicago: Encyclopedia Britannica, 1952), s.v. "Justice." There is also a new anthology, edited by Robert Solomon and Mark Murphy, *What Is Justice?* (New York: Oxford University Press, 1990), that includes both classical and contemporary texts.

5. Studs Terkel, *Working: People Talk about What They Do All Day and How They Feel about What They Do* (New York: Random House, 1972), xxix-xxx.

6. See note 1, above.

7. Byrne, *Work, Inc.*, 3.

8. Ibid., 277.

9. Ibid., 281.

10. See note 4, above.

11. Byrne, *Work, Inc.*, 45.

12. Johan Huizinga, *Homo Ludens: A Study of the Play Element in Culture* (Boston: Beacon, 1967).

13. Byrne, *Work, Inc.*, 99 and 109.

14. Ibid., 115.

15. Ibid., 120.

16. Ibid., 135.

17. Ibid., 17.

18. Ibid., 212.

19. Ibid., 218.

20. Ibid.

21. Judith Lichtenberg, "Workers, Owners and Factory Closings," *Report from the Center for Philosophy and Public Affairs* (Fall 1984):12.

22. Terkel, *Working*, 264-265.

23. Michael W. McCann, *Taking Reform Seriously: Perspectives on Public Interest Liberalism* (Ithaca, N.Y.: Cornell University Press, 1986), 262.

Chapter 14. Conservatives (pages 158-169)

1. E. Digby Baltzell, *Puritan Boston and Quaker Philadelphia: Two Protestant Ethics and the Spirit of Class Authority and Leadership* (New York: Free Press, 1979), 28.

2. Ibid., 30.

3. Ibid., 507.

4. Carl Cohen, *Four Systems* (New York: Random House, 1982).

5. Russell Kirk, *The Conservative Mind* (Chicago: Regnery, 1953), and, as editor, *The Portable Conservative Reader* (New York: Viking/Penguin, 1982).

6. See Peter Viereck, "Conservatism," in *The New Encyclopedia Britannica* (Chicago: Encyclopedia Britannica, 1974), *Macropaedia*, vol. 5, 62-69.

7. See Kenneth Minogue, "Conservatism," in *The Encyclopedia of Philosophy* (New York: Free Press and Macmillan, 1967), vol. 2, 195-198.

8. John M. Finnis, *Natural Law and Natural Rights* (Oxford: Oxford University Press, 1980).

9. John Courtney Murray, *We Hold These Truths* (New York: Sheed and Ward, 1960).

10. Thomas Aquinas, *Summa theologiae*, I-II, 90-97.

11. Finnis, *Natural Law*, 85-86.

12. Murray, *We Hold These Truths*, 325.

13. Ibid., ix.

14. Ibid., 335.

15. Ibid., 13-14.

16. Ibid., 179.

17. Ibid., 200.

18. Both of the cited documents can be found in *The Documents of Vatican II*, ed. W. Abbott (New York: Guild Press, 1966), as can "The Church in the Modern World," discussed at length in the next four paragraphs. It is customary to reference official Roman Catholic documents by numbered paragraphs rather than by page numbers.

19. "The Church in the Modern World," no. 5.

20. Ibid., nos. 2 and 3.

21. This papal letter and others cited in the text are available in D. O'Brien and T. Shannon, eds., *Renewing the Earth: Catholic Documents on Peace, Justice, and Liberation* (Garden City, N.Y.: Doubleday, 1977). See also Joseph Gremillion, ed., *The Gospel of Peace and Justice: Catholic Social Teaching since Pope John* (Maryknoll, N.Y.: Orbis Books, 1976).

22. Germain Grisez, *Contraception and the Natural Law* (Milwaukee, Wis.: Bruce, 1964); see also Grisez and Russell Shaw, *Beyond the New Morality* (Notre Dame, Ind.: University of Notre Dame Press, 1974).

23. Kirk, *The Conservative Mind*, 542.

24. Paul T. Durbin, "Thomism and Technology: Natural Law Theory and the Problems of a Technological Society," in *Theology and Technology*, ed. C. Mitcham and J. Grote (Lanham, Md.: University Press of America, 1984), 209-225.

Chapter 15. Liberals: Self-Interested and Progressive (pages 170-188)

1. See Daniel Bell, *The Cultural Contradictions of Capitalism* (New York: Basic Books, 1978), 3.

2. Ibid., xi.

3. Carl Cohen, *Four Systems* (New York: Random House, 1982), 67 and 34.

4. See F. Bolkestein, ed., *Liberalism: Conversations with Liberal Politicians* (New York: Elsevier, 1982), 281.

5. Bell, *Cultural Contradictions*, xiv.

6. John Kenneth Galbraith, *Economics and the Public Purpose* (Boston: Houghton Mifflin, 1973), 236 and 237.

7. Edward Walter, *The Immorality of Limiting Growth* (Albany: State University of New York Press, 1981).

8. Ibid., 3-6.

9. Ibid., 61.

10. Robert Nozick, *Anarchy, State, and Utopia* (New York: Basic Books, 1974), xxx.

11. Bernard Gendron, *Technology and the Human Condition* (New York: St. Martin's, 1977), 14.

12. Ronald Dworkin, *Taking Rights Seriously* (Cambridge, Mass.: Harvard University Press, 1977), 264-265.

13. Marshall Cohen quoted in N. Daniels, ed., *Reading Rawls* (New York: Basic Books, 1974), xiv.

14. John Rawls, *A Theory of Justice* (Cambridge, Mass.: Belknap/Harvard University Press, 1971), 242-243.

15. Ibid.

16. Ibid., 190.

17. Ibid., 191.

18. Walter, *Limiting Growth*, 161.

19. Galbraith, *Economics and the Public Purpose*, 239.

20. Ibid., 240-251.

21. Thomas Nagel, *The Possibility of Altruism* (Oxford: Clarendon/Oxford University Press, 1970), 15.

22. Ibid., 145-146.

23. Nicholas Rescher, *Unselfishness: The Role of the Vicarious Affects in Moral Philosophy and Social Theory* (Pittsburgh, Pa.: University of Pittsburgh Press, 1975), ix.

24. Ibid., 100.

25. Ibid., 102.

26. Ibid., 104.

27. But see Karen Hanson, *The Self Imagined: Philosophical Reflections on the Social Character of Psyche* (New York: Routledge and Kegan Paul, 1986).

28. See R. W. Sleeper, *The Necessity of Pragmatism: John Dewey's Conception of Philosophy* (New Haven, Conn.: Yale University Press, 1986), chapter 7; and Hans Joas, *G. H. Mead: A Contemporary Re-Examination of His Thought* (Cambridge, Mass.: MIT Press, 1985), chapter 6.

29. William James had spelled out the principle involved here before Mead and Dewey. See his "The Moral Philosopher and the Moral Life," reprinted in J. McDermott, *The Writings of William James: A Comprehensive Edition* (New York: Random House, 1967), 610-629.

30. Rawls, *Theory of Justice*, 190-191.

31. For his fullest and most recent formulation, see Hans Jonas, *The Imperative of Responsibility: In Search of an Ethics for the Technological Age* (Chicago: University of Chicago Press, 1984).

32. For Dewey, see Sleeper, *Necessity of Pragmatism*, and Larry A. Hickman, *John Dewey's Pragmatic Technology* (Bloomington: Indiana University Press, 1990). For Mead, see Joas, *G. H. Mead*.

33. See Cornel West, *The American Evasion of Philosophy: A Genealogy of Pragmatism* (Madison: University of Wisconsin Press, 1989), especially references throughout the book to the work of Richard Rorty.

Chapter 16. Academic Philosophers (pages 189-201)

1. George Allan, *The Realizations of the Future: An Inquiry into the Authority of Praxis* (Albany: State University of New York Press, 1990).

2. Thomas Perry, *Professional Philosophy: What It Is and Why It Matters* (Dordrecht: Reidel, 1986), xiii.

3. Ibid.

4. Ibid., xiv.

5. Ibid., xiii.

6. W. V. Quine, *Philosophy of Logic* (Englewood Cliffs, N.J.: Prentice-Hall, 1970), 4-5.

7. Bertrand Russell, *A History of Western Philosophy* (New York: Simon and Schuster, 1945), 834.

8. Hans Reichenbach, *The Rise of Scientific Philosophy* (Berkeley: University of California Press, 1951), 123-124.
9. Herbert Marcuse, *One-Dimensional Man* (Boston: Beacon Press, 1964), 182 (as just one example).
10. See Jacques Ellul, *The Technological Society* (New York: Knopf, 1964), for instance, 419 and 432-436.
11. Richard Rorty, *Philosophy and the Mirror of Nature* (Princeton, N.J.: Princeton University Press, 1979), and *The Consequences of Pragmatism* (Minneapolis: University of Minnesota Press, 1982). See also Rorty's *Contingency, Irony, and Solidarity* (New York: Cambridge University Press, 1989), as well as Konstantin Kolenda, *Rorty's Humanistic Pragmatism: Philosophy Democratized* (Tampa: University of South Florida Press, 1990).
12. John Dewey, *Reconstruction in Philosophy*, 2d ed. (Boston: Beacon Press, 1948), xi-xii.
13. John Dewey, *Liberalism and Social Action* (New York: Putnam, 1935), 92.
14. G. H. Mead, "Suggestions toward a Theory of the Philosophical Disciplines," in *Selected Writings*, ed. A. Reck, (Indianapolis, Ind.: Bobbs-Merrill, 1964), 6-24, esp. 20-22.
15. Mead, "Scientific Method and the Moral Sciences," in *Selected Writings*, 266.
16. See William James, "The Moral Philosopher and the Moral Life," reprinted in McDermott, ed., *The Writings of William James: A Comprehensive Edition* (New York: Random House, 1967), 610-629; see esp. 624-625.
17. See Gabriel A. Almond, Marvin Chodorow, and Roy Harvey Pearce, eds., *Progress and Its Discontents* (Berkeley: University of California Press, 1982), and Steven L. Goldman, ed., *Science, Technology, and Social Progress* (Bethlehem, Pa.: Lehigh University Press, 1989).
18. See Thomas S. Kuhn, *The Structure of Scientific Revolutions*, 2d ed. (Chicago: University of Chicago Press, 1970), 176.
19. G. H. Mead, *Philosophy of the Act* (Chicago: University of Chicago Press, 1938), 65.
20. Mead, "Scientific Method and Individual Thinker," in *Selected Writings*, 171-211.
21. Ibid., 196.
22. Ibid., 203.
23. Kuhn, *Structure of Scientific Revolutions*, 8.
24. Ibid.
25. Kuhn, "Reflections on My Critics," in *Criticism and the Growth of Knowledge*, ed. I. Lakatos and A. Musgrave (Cambridge: Cambridge University Press, 1970), 272-273.
26. Paul K. Feyerabend, *Against Method* (London: NLB, 1975).
27. John Ziman, *An Introduction to Science Studies: The Philosophical and Social Aspects of Science and Technology* (Cambridge: Cambridge University Press, 1984), 110; the other reference is to Ludwig Fleck, *Genesis and Development of a Scientific Fact* (Chicago: University of Chicago Press, 1979).
28. Ziman, *Introduction to Science Studies*, 112.
29. Ibid.
30. Rom Harré, preface to Karin Knorr-Cetina, *The Manufacture of Knowledge: An Essay on the Constructivist and Contextual Nature of Science* (Oxford: Pergamon, 1981), viii.
31. Bruno Latour and Steve Woolgar, *Laboratory Life: The Social Construction of Scientific Facts* (Beverly Hills, Calif.: Sage, 1979). For a similar approach to technology, see W. Bijker, T. Hughes, and T. Pinch, eds., *The Social Construction of Technological Systems* (Cambridge, Mass.: MIT Press, 1987).
32. Joseph Margolis, *Pragmatism without Foundations: Reconciling Realism and Relativism* (New York: Blackwell, 1986).
33. Ibid., 10.
34. Ibid., 26.
35. Ibid.

36. Ibid., 27.
37. Ibid.
38. Ibid., 209.
39. Ibid.
40. Ibid., 208.
41. Ibid.
42. Ibid.
43. John Dewey, *The Quest for Certainty* (New York: Minton, Balch, 1929).
44. Cornel West, *The American Evasion of Philosophy: A Genealogy of Pragmatism* (Madison: University of Wisconsin Press, 1989).
45. Margolis, *Pragmatism without Foundations*, 209.
46. Ibid.
47. See, for example, Ralph W. Sleeper, *The Necessity of Pragmatism: John Dewey's Conception of Philosophy* (New Haven, Conn.: Yale University Press, 1986); Thomas A. Alexander, *John Dewey's Theory of Art, Experience and Nature* (Albany: State University of New York Press, 1987); and Larry A. Hickman, *John Dewey's Pragmatic Technology* (Bloomington: Indiana University Press, 1990).
48. Albert Borgmann, *Technology and the Character of Contemporary Life* (Chicago: University of Chicago Press, 1984), and *Crossing the Postmodern Divide* (Chicago: University of Chicago Press, 1992).
49. Bruce Kuklick, *The Rise of American Philosophy: Cambridge, Massachusetts 1860-1930* (New Haven, Conn.: Yale University Press, 1977).

Index